Science and Technology
in the Early Years

OPEN UNIVERSITY PRESS
Gender and Education Series

Editors
ROSEMARY DEEM
*Senior Lecturer in the School of Education at the
Open University*
GABY WEINER
*Principal Lecturer in Education at
South Bank Polytechnic*

The series provides compact and clear accounts of relevant research and practice in the field of gender and education. It is aimed at trainee and practising teachers, and parents and others with an educational interest in ending gender inequality. All age-ranges will be included, and there will be an emphasis on ethnicity as well as gender. Series authors are all established educational practitioners or researchers.

TITLES IN THE SERIES

Boys Don't Cry
Sue Askew and Carol Ross

Science and Technology in the Early Years
Naima Browne (ed.)

Untying the Apron Strings
Naima Browne and Pauline France (eds)

Changing Perspectives on Gender
Helen Burchell and Val Millman (eds)

Co-education Reconsidered
Rosemary Deem (ed.)

Women Teachers
Hilary de Lyon and Frances Widdowson Migniuolo (eds)

Girls and Sexuality
Lesley Holly (ed.)

A History of Women's Education in England
June Purvis

Whatever Happens to Little Women?
Christine Skelton (ed.)

Dolls and Dungarees
Eva Tutchell (ed.)

Just a Bunch of Girls
Gaby Weiner (ed.)

Women and Training
Ann Wickham

Science and Technology in the Early Years

AN EQUAL OPPORTUNITIES APPROACH

Edited by Naima Browne

Open University Press
Milton Keynes · Philadelphia

Open University Press
Celtic Court
22 Ballmoor
Buckingham
MK18 1XW

and
1900 Frost Road, Suite 101
Bristol, PA 19007, USA

First Published 1991
Reprinted 1995

British Library Cataloguing in Publication Data

Science and technology in the early years: an equal
   opportunities approach. — (Gender and education).
   1. Great Britain. Primary schools. Students. Girls.
   Curriculum subjects. Science. Equality of opportunity
   I. Browne, Naima   II. Series
   372.350941

   ISBN 0-335-09229-2

Library of Congress Cataloging-in-Publication Data

Science and technology in the early years: an equal opportunities
   approach/edited by Naima Browne.
         p.   cm. — (Gender and education series)
   Includes bibliographical references.
   Includes index.
   ISBN 0-335-09229-2
      1. Science — Study and teaching (Preschool)   2. Sex differences
   in education.   3. Educational equalization.   I. Browne, Naima.
   II. Series.
   LB1140.5.S35S38   1990
   372.3'5—dc20
                                                        90-7637 CIP

Typeset by Colset (Pte.) Limited, Singapore
Printed in Great Britain by J. W. Arrowsmith Limited, Bristol

Contents

SECTION 2: SCHOOL ACCOUNTS AND ACTION

List of Contributors

Kim Beat has been a primary teacher in the London Borough of Brent for twelve years, and was an Advisory Teacher for Equality in Education for two years. She is currently the deputy head at Barham School, Wembley. She has been a member of the 'Design It, Build It, Use It Collective' along with Julie Cahill, Jan McAleavey, Uma Pandya, Myra Joyce and Tina Worrall. The collective have written a booklet and run a number of workshops.

Naima Browne taught for several years in Inner London schools. She has worked as an early-years advisory teacher for both bilingualism and equal opportunities. She was recently a senior lecturer in early-years education at South Bank Polytechnic and is currently lecturing at Goldsmith's College, London. Her research interests include a range of issues connected with equal opportunities in early-years education. She has just completed her doctoral research into the educational provision made for very young children in nineteenth-century London.

Julie Cahill has been teaching in Brent since the early 1980s. She has taught across the whole nursery and primary age range but has mainly taught infants. She is a member of the 'Design It, Build It, Use It Collective'. As Child Protection Advisory Teacher in Brent she is concerned with issues relating to child abuse awareness and protection approached from an equal opportunities perspective.

Eve Lyon is currently head of Abbey Wood Nursery School in South London.

Patience MacGregor was born in London. In the 1970s she worked as a tutor in adult education, in the 1980s as an early-years teacher in Lambeth and Tower Hamlets (ILEA), and from 1986 to 1988 as an early-years advisory teacher. She is currently teaching at Rachel Macmillan Nursery School, Deptford, and is supporting parent, school and community links.

Uma Pandya has taught in ILEA and Brent for seven years; she has taught children aged from 3 to 11. She is currently completing an MA in Curriculum Studies at the London University Institute of Education. As an Advisory Teacher for equal opportunities in Brent she was involved in the Girls in Construction Project with 13- and 14-year-old girls in secondary school.

Rose Parkin lives in East London where she has been teaching primary and pre-school children since 1977. She was involved in setting up a co-operative nursery and co-ordinated its work for two years. Following this she became mathematics co-ordinator in a small primary school where much of her work for this chapter was carried out. She now works in a large infants' school where she is responsible for science, computers, AVA and CDT. She has two young children.

Carol Ross has been working in the field of Equal Opportunities in Education for the past seven years. Currently she is an Equal Opportunities Advisory Teacher for Islington LEA and is lecturing in gender and education at the London University Institute of Education.

Jane Savage has taught across the whole primary age range and has been both an advisory teacher for primary science and a deputy head teacher in ILEA. She is currently working at the Institute of Education, London University as a primary education lecturer with responsibility for primary science. Her interests are in the fields of primary planning and assessment and classroom organization.

Iram Siraj-Blatchford is a lecturer in early-years education at the University of Warwick faculty of Eductional Studies.

Claudette Williams has been a primary teacher for over ten years. She started teaching in Santley Infant School, in South London. On Saturdays she worked with the Ahfiwe Supplementary School, providing Black children with an opportunity to gain insight into the contribution their parents and other Black people have made to British Society.

In 1984 she was seconded to the Afro-Caribbean Education Resource Team (ACER) to develop multi-cultural/anti-racist materials for use in primary schools. On completion of her year at ACER she joined a dynamic team of ten early-years teachers in the Bilingual Under Fives Team (BUF) at the Centre for Urban Educational Studies. It offered support to nursery teams working with bilingual children throughout the ten divisions of ILEA. On return in 1989 from a one-year teacher exchange to Trinidad, she joined the School of Teaching Studies at the Polytechnic of North London.

Barbara Wyvill is a Science Consultant specializing in anti-racist and anti-sexist science education at home and abroad. When she left school, her Headmistress told Barbara that she would make someone a very good wife: she did, and had three children, so she became a primary-school teacher. Transferring to a secondary school she specialized in science and did a part-time M.Sc. in entomology. She graduated to in-service teacher education, first with the JISTT project and subsequently as director of the North London Science Centre.

Series Editor's Introduction

Education in the early years, i.e. 3 to 8, was somewhat neglected in the early 1980s by teachers and researchers with an interest in gender issues. This stems both from the apparent invisibility of gender difference in the early years of schooling – there are no examinations or youth cultures to draw attention to inequality – and the relatively low status of teaching and researching younger children in the schooling system.

The Gender and Education series is attempting to fill this vacuum by including a number of volumes on primary schooling and early-years education. This latest collection is an important addition to earlier discussions in the two primary volumes, *Whatever Happens to Little Women?* and *Dolls and Dungarees*. It should, however, also be seen as the sequel to the only other volume in the series focusing specifically on early-years education *Untying the Apron Strings* also edited by Naima Browne (with Pauline France).

This volume is important because it is the first to address gender issues in science and technology in early-years education. Whereas gender divisions within these subjects have received much attention at secondary level, (e.g. Catton, 1985; DES, 1980; Harding, 1983; Kelly, 1987; Millman, 1984; Smail, 1984; Whyte, 1986), few have thought them relevant to the educational experiences of younger children. However, with the advent of the National Curriculum* which includes science as one of the three core subjects, and technology as one of the seven foundation subjects, preparation for the study of these subjects in the early years of schooling can now be perceived as crucial.

This volume has been written by women from a variety of

backgrounds in early-years education – nursery and infant teachers, heads, advisers and lecturers – and from a range of ethnic and class backgrounds. Readers, themselves probably teachers, student teachers or related professionals, are thus able to draw on the work and critical perspectives of experienced fellow (if that's the appropriate word!) practitioners in the improvement of their own practice. It highlights current experiences of girls in schools and students on initial training courses and then moves on to give practical examples of curriculum developments and teacher strategies which have proved fruitful. It also shows developments in thinking along the continuum from equal opportunities to anti-sexist practice and how good early-years practice is not necessarily incompatible with the demands of the National Curriculum.

At a time when the equal opportunities outcomes of the educational reforms depend much on the energy, creativity, commitment and skills of practising teachers, it is hoped that this collection will be of considerable help to them (and others) in promoting educational equality in the 1990s.

Gaby Weiner

Note

* The National Curriculum applies to only England and Wales, and not to Scotland and Northern Ireland.

References

Catton, J. (1985). *Ways and Means: the Craft, Design and Technology Education of Girls*, York, Longman.

Department of Education and Science (1980). *Girls and Science*, HMI Series, *Matters for Discussion* 13, London, HMSO.

Harding, J. (1983). *Switched Off: the Science Education of Girls*, York, Longman.

Kelly, A. (ed.) (1987). *Science for Girls*, Milton Keynes, Open University Press.

Millman, V. (1984). *Teaching Technology to Girls*, Coventry, Elm Bank Teachers' Centre.

Smail, B. (1984). *Girl Friendly Science: Avoiding Sex Bias in the Curriculum*, York, Longman.
Whyte, J. (1986). *Girls into Science and Technology*, London, Routledge & Kegan Paul.

Acknowledgements

I should like to thank the following people for their help with this book: the Ahmeds, Rosemary Deem, Gaby Weiner and all the contributors who have worked so hard. Finally, I should also like to thank Phil Browne.

The Present Situation

CHAPTER 1

Science and Technology in the Early Years of Schooling: An Introduction

NAIMA BROWNE

Most early-years practitioners would agree that the early years of schooling are crucial in the development of attitudes, skills and concepts. This phase of education, however, has tended to be neglected when considering educational inequalities in science and technology. Money for research has tended to be directed towards examining the science education of older girls rather than the very young. The low involvement of girls in science and technology is not an exclusive feature of secondary or junior schools. Gender differences in these curriculum areas are strikingly evident amongst girls and boys in nurseries and infant schools. Many early-years teachers have been uncertain about the possible causes of these differences and have been unsure about what strategies are likely to be most effective in minimizing these inequalities. One reason for this uncertainty is that most of the widely available published research into girls' science education has tended to focus on older children who work in classrooms in which the ethos and organization is very different from that of nursery and infant schools. This book raises some of the important issues and provides early-years teachers with tangible ideas about how they can begin to work towards ensuring that all children are enabled to achieve their full potential in science and technology.

The authors are involved in early-years education, either as teachers, advisers or early-years lecturers involved in initial teacher education. All the writers emphasize the importance of a child-centred approach to education in the early years (i.e. an approach which builds on the interests, strengths and needs of individual children). The writers, however, have different views about the reasons for and implications of the low levels of involvement by girls in science. Some of the authors adopt an equal opportunities approach that focuses on improving girls' access to the existing curriculum and scientific community. This approach can be defended on the grounds that it is important that girls have access to current scientific understandings and know the rules by which the scientific community operates. In addition, it is important that more girls and women become members of the scientific community in order that women's voices are heard. Other authors have adopted a more radical approach and argue for anti-sexist and anti-racist science education. This approach is also defensible, especially if one believes that science is not value-free and objective. Those adopting an anti- sexist, anti-racist approach focus on the curriculum itself and question its validity. They argue that the existing science curriculum initiates children into one way of thinking about science, it does not help children become aware of or value the range of scientific methods and interpretations. Furthermore, those adopting the latter approach question whether an increase in the proportion of women scientists is sufficient to bring about a change in the power relations within the scientific community. The authors' classroom strategies for minimizing inequalities reflect their individual perspective.

This book has been organized to form two sections. The first five chapters explore the current situation in schools and initial teacher education establishments and highlight what needs to be done to overcome inequalities in the science education of young children. The following six chapters provide accounts of the steps taken by schools and individual teachers in an effort to ensure that all children in early-years classes have the chance to develop to their full potential in science. Two chapters focus on resources and assessment. The final chapter outlines INSET strategies which have proved successful in supporting early-years teachers who are keen to teach science but who, because of past experiences, are unsure about their own skills.

The introduction of the National Curriculum has meant that science and technology now form a major part of the early-years curriculum. It could be argued that this is a positive step since our daily lives are influenced by scientific and technological developments. In order to function effectively in present-day society, girls and women need to be scientifically and technologically literate. Scientists, the majority of whom are male, put forward conflicting arguments about such issues as the benefits of certain kinds of medical treatment, the safety or irradiated food, the causes and effects of environmental pollution, etc. Those members of society without basic scientific understandings are not well equipped to make their own decisions about the validity of the various arguments, methods of investigations and interpretations and, instead, must adopt the role of bystander as others make decisions which could have a profound effect on their lives. It is important that girls are enabled to achieve their full potential in science not simply in order to increase the proportion of female professional scientists but also to help increase the control that women and girls have over their own lives. The articles in this book offer some practical suggestions for ways of helping girls achieve this autonomy.

CHAPTER 2

The Ideological Context of Science Education in the Early Years: An Historical Perspective

NAIMA BROWNE

Introduction

This chapter places the science education of very young girls in its social, political and historical context by tracing how early-years science education has changed since infant schools were first established in Britain. In so doing it demonstrates that ideas about what constitutes 'good' early-years science education change to accommodate the changing views about the nature and purpose of early-years education in general.

The early-years curriculum has not remained static since Robert Owen opened his pioneering school for infants in 1816. This is hardly surprising when one considers how British society has changed over the last two centuries. At any point in time schooling is influenced by the society within which it takes place. An historical analysis of the early-years curriculum shows clearly how, in addition to political and economic factors, it has been moulded by a range of factors including the views current in wider society about the relative value and status of different types of knowledge and skills, views about the aims of early-years education and the adoption of different learning theories.

In the nineteenth century the schooling a child received was dependent upon the child's social class and gender. Most working-class children attended the monitorial schools or elementary schools in which the teaching was mechanical and the curriculum narrow. The three Rs were taught but religious and moral education formed a major part of the curriculum in the hope that children from poor homes would be educated to accept their allotted station in life and become docile workers who would not challenge the existing social status quo.

Those arguing for a wider curriculum for working-class infants had to be careful to stress that, as a consequence, the children's moral and religious education would not suffer. The Bishop of London was in favour of broadening the curriculum to include 'certain portions of history, geography, [and] the elements of useful, practical science'. In order to minimize opposition to his ideas from those who believed that the children's moral education would suffer he added that 'it is found that the children are not at all less interested in the religious part of their education, in consequence of their attention being occasionally directed to other branches of learning' (PP, 1834).

Only a few of those who argued for a wider curriculum did so from a belief that knowledge would empower individuals and enable them to bring about changes in society. One writer for example asked 'Why should some knowledge which can be acquired by all remain in the possession of the privileged few?' but made it clear that he was not arguing for equality when he added that 'a man with an open mind . . . has much greater chance of understanding and fulfilling the duties of his station, than if brought up in gross ignorance' (*Quarterly Journal of Education.* 1833).

The education of working-class girls and boys in publicly funded schools was similar until they reached the age of 6 years. Thereafter the education that girls received was usually not as extensive as that of boys. One reason for this was that girls spent most afternoons sewing whilst the boys had additional lessons. The deleterious effect of this on girls' attainment was commented on by a school inspector:

> With regard to girls' schools it certainly seems a hardship to expect girls (*who are generally supposed to have less aptitude for arithmetic than boys,* and who are employed for nearly half the day in sewing) to do the same sums as the boys. (PP, 1867: xxxiii, 32)

Many parents also felt that the education of boys was more important than that of girls and this was reflected in the school attendance figures for girls and boys. In the 1830s girls accounted for only 30 per cent of the children on the registers of London schools associated with the British and Foreign School Society. The explanation given by the Society was that 'parents make the mistake of thinking that reading and writing are not so important for girls' (*Quarterly Journal of Education*, 1833). Many girls were required to help at home which was yet another reason why girls suffered educationally. The school life of both girls and boys was often very short as schooling was expensive and many were required to start earning at a young age. It is not hard to understand the parents' concern that their children be taught the basic skills and that if, in the case of girls, parents believed that reading and writing were not thought to be very important it was unlikely that they would have pressed to have their daughters taught natural history or simple mechanics during their short school career.

It is against this background of publicly funded education that developments in science education in the early years need to be viewed.

Learning through activity and the use of the senses: theory and reality in the nineteenth century

The inclusion of science in the curriculum for very young children is not something new. In the nineteenth century forward-looking educationists were making moves to extend the curriculum for young children in infant and elementary schools beyond the four Rs (R.E., reading, writing, and 'rithmetic) to include other subjects, of which natural history was one.

Robert Owen had included science (or natural history) in his infant-school curriculum in 1816 because of his belief that working-class children would benefit from a wide curriculum. His infant school in New Lanark was attended by the children of the workers in his cotton mills and the curriculum consisted of reading, writing, arithmetic, sewing, geography, history, R.E., singing, dancing and natural history. The learning environment was designed to interest and stimulate the children and the Pestalozzian object lesson was used as a method of teaching:

> The schoolroom was furnished with paintings, chiefly of animals, with maps and often supplied with natural objects from the gardens, fields and woods – the examination and explanation of which

always excited their curiosity and created animated conversation between the children and their instructors. (Whitbread, 1972)

From a late twentieth-century standpoint there may not appear to be anything unusual about Owen's curriculum and, as regards the children's exploration of the natural world, it could be argued that in Owen's schools this was very limited and in no way compares to that which currently occurs in good nursery and infant classes. It is important, however, to view the New Lanark school in its historical context. Whilst the children in New Lanark were being given the opportunity to handle objects and talk about them, numerous children of the same age were being taught in the monitorial schools.

The New Lanark school attracted attention and interest and gradually schools specifically for infants began to be established throughout the country. One such school was in Spitalfields, London, where Samuel Wilderspin was the teacher. Wilderspin became a very influential figure in the development of infant schools in Britain. Wilderspin, like Owen, advocated a wide curriculum for working-class children. His aim was that by the time children left an infant school they should, amongst other things, 'have a tolerable knowledge of the quality of such things as immediately come under its notice' and 'a slight knowledge of the elements of natural history, of the habits and manners of different animals' (PP, 1835). He claimed that the main aim of an infant school was to 'set the children thinking' and, through the use of pictures and objects he hoped that the children would be encouraged to observe, compare and contrast, articulate what they see and test any inferences they may have drawn. In theory at least it would appear that the children developed a number of scientific skills.

His practice, however, did not reflect his theories. Much of Wilderspin's teaching, despite his efforts to make it interesting and concrete, relied on rote-learning. This may have been due in part, to a lack of understanding of the theories relating to the importance of first-hand experience and in part to Wilderspin's belief that two teachers were sufficient for 200 2- to 6-year-olds! When the 1835 Education Select Committee asked Wilderspin to explain how he taught natural history his reply revealed just how mechanical, pedantic and stultifying the teaching in infant schools could be:

We assemble them at one end of the school . . . we would put a picture of a horse before the children and we would say, 'What is this?' they would say 'It is a horse'. 'No, you are mistaken, this is not a horse.' 'Yes it is', some of the positive ones would say: 'No it's not, you do not think.' Then others would say, 'It is a picture of a horse', and we should say, 'Now you are right; it is a representation of a horse but not a horse itself.' Now we go into the parts of the horse; we teach them to understand which are the pasterns, and the fetlocks, the shanks, the withers and the croups . . . and so on. . . . Then we come to the uses of the animal. . . . Then we would say 'What is made of the skin?' . . . and so on.

The overemphasis on facts and the underemphasis on drawing on the children's experiences and knowledge was highlighted by Dickens. In *Hard Times* young Sissy Jupe, whose father was a horse breaker, was unable to define a horse when she was asked to do so and it fell to a boy in the class to furnish the required facts, 'Quadriped. Graminovourous. Forty teeth, namely twenty-four grinders, four eye teeth and twelve incisive . . .' (Dickens, 1854).

The 'object lesson', used by both Owen and Wilderspin, played a significant part in the science education of young children throughout the nineteenth century. The idea was originally Pestalozzi's who emphasized the important role of observation and verbalization in conceptual development. He eschewed rote-learning and insisted that children ought to use their senses to discover as much as possible about an object before being told about its qualities, uses and origin.

Elizabeth and Charles Mayo were responsible for disseminating many of Pestalozzi's ideas, particularly that of the object lesson. In 1829, Elizabeth Mayo's book, *Lessons on Objects*, was widely used and went through 16 editions in 30 years. The book consisted of descriptions of common objects such as glass, leather and bread and listed the salient features of the objects in question.

The system of object lessons received official recognition. HMIs who visited infant schools in receipt of government grants were required to ascertain whether 'knowledge of natural objects' was included in the curriculum, whether the school possessed a cabinet containing natural objects and also whether the children were 'exercised in examining and describing in very simple and familiar terms the properties of the Natural Objects by which they are surrounded' (Rusk, 1933).

Unfortunately, many teachers who used object lessons had no understanding of the underlying principles and the lessons degenerated into mechanical rote-learning of facts. A contemporary critic stated that he had never seen a teacher use Elizabeth Mayo's book without finding that:

> . . . the little pupils during the greater part of their lesson were not really learning the properties of glass, or chalk, or copper wire, but were in fact learning the meaning of sundry hard words such as *transparent, opake, friable, malleable, ductile, insipid, sapid.* . . . (Malden, 1838)

The fault did not lie totally with the teachers but rather with an overestimation of what it is possible to discover through observation alone. The section on Mayo's book devoted to bread suggested that as a result of close observation the children would discover various qualities of bread (e.g. 'that it is absorbent, opaque, solid, edible [!] (E. Mayo, 1837). It is hard to imagine how observation of a loaf of bread would lead a child to find out that bread is absorbent or what absorbent means!

The method remained popular throughout the nineteenth century and was the main means by which natural history was taught to young children. Object lessons may have been popular because the method did not require that the teacher possessed any scientific knowledge or understanding. Few of the teachers would have possessed much scientific knowledge as many had themselves been educated in monitorial or elementary schools and it was only after the middle of the nineteenth century that it became compulsory for student teachers to study some science on their training course.

Many young working-class children at school received no scientific education of any description. One reason for this could have been that many of the teachers had themselves received no training in science education. It was reported, for example, that none of the natural sciences were allowed to be taught in the National Society's Central Model School in Westminster and yet this school was used in the training of teachers (Central Society of Education, 1838).

In other schools the school managers specifically banned teaching beyond the three Rs. The result in some schools was that:

> no instruction is ever given in the elements of any of the sciences. . . . A master does not teach a class that the earth moves

round the sun, and that the sun does not move round the earth; nor does he quicken the pupil's inventive faculties by teaching him the principle of the mechanical powers. (Central Society of Education, 1838)

In terms of science education, working-class children who attended private working-class schools (e.g. dame schools and common day schools) rarely fared better than their contemporaries in publicly funded schools. The curriculum was usually limited to the three Rs and, for the girls, sewing and knitting. In some common day schools, the curriculum was extended to include history and geography. In order to stay in business teachers in private working-class schools had to be sensitive to the demands of their clients, and since most working-class parents demanded the three Rs the schools responded accordingly.

Social class and the science education of young children

Middle-class and working-class children had very different educational experiences. The curriculum in schools attended by middle-class children tended to be broader and the teachers were better educated than those teaching working-class children. A brief account of the teaching at the Bruce Castle School, Tottenham, illustrates clearly how the science education of young middle-class children differed to that of working-class children of the same age.

Bruce Castle was attended by children aged between 4 and 9. One of the main principles of the school was that children were to be taught through the use of their senses and 'facts and things' were to be the medium and the object of learning. Furthermore, it was claimed that 'the acquirement of a large quantity of information [was] considered but a secondary object, when compared to a thorough comprehension of what [was] learnt, and to the habit of attentive inquiry and investigations' (Central Society of Education, 1838). To support their aims the school possessed a museum which contained a range of specimens drawn from both the natural world and the world of art. A contemporary noted that various insects and small creatures were eagerly observed by the children, lectures were delivered on the simplest principles of mechanics and useful machines were 'dissected' and the workings explained (Central Society of Education, 1838). The school also believed that the children's understanding of the natural world would be further stimulated through the reading and recitation of relevant stories and poems.

It would seem that a middle-class girl attending the Bruce Castle School had more in common with an infant-school child of the 1980s than with one of her working-class contemporaries in a nineteenth-century infant school. The use of lectures was typical of the last century but the negative effect of these was outweighed by 'modern' practices such as the emphasis on process and under-standing rather than information, allowing children the freedom to explore their environment and an awareness of the value of cross-curricular links.

Girls as well as boys attended Bruce Castle School but not all middle-class girls were as fortunate when it came to science education. In the nineteenth century demands were being made for a better education for middle-class girls. Those arguing for such improvements in girls' education did not necessarily envisage that girls and boys would receive the same education. The dis-cussions focused on the education of older girls but obviously there were implications for younger girls. Mrs Mudie, an experienced teacher, stated that 'The great point at which female education should aim, is the communicating to ladies as much of the general principles of knowledge as shall make it not rude to talk to them upon any ordinary subject of a literary or scientific nature.' In relation to science, it would seem that Mudie felt that girls needed to be taught just enough to ensure the girls would become social assets but not so much as to enable them to engage in scientific investigations as adults! Mrs Sandford, another female writer on middle-class girls' education, did girls few favours when she stated that she was not sure that it was wise to encourage girls to engage in scientific study as she believed them constitutionally more suited to literary studies (Kamm, 1965). The attendance figures for the British Association for the Advancement of Science (BAAS) suggested that many women did not share these views. In the 1830s, women often outnumbered men at BAAS meetings. Attendance at meetings, however, did not mean that women scientists were taken seriously by the community of male scientists. Some women, however, were still able to pursue scientific investigations and contributed greatly to the field of science. One such woman was Mary Somerville who argued 'for the emancipation of my sex from the unreasonable prejudice too prevalent in Great Britain against literary and scientific education for women' (Alic, 1986).

Between the 1860s and the 1890s educational facilities for middle-class girls improved, not only in terms of quantity but also in that the curriculum was broadened and included subjects previously taught only to boys.

In the middle of the nineteenth century there was an increased interest in the teaching of science fuelled in part by the fear that if Britain was to fall behind its European competitors in the fields of science and technology, the economic consequences would be dire. In elementary schools, science teaching was influenced by the views of Dawes who placed the emphasis on knowledge and argued that scientific knowledge was useful knowledge in that children would benefit from understanding the natural world as scientists understood it. When discussing the education of middle-class children the emphasis was different as was revealed in the Taunton Committee's statement on science education:

> True teaching of science consists not merely of imparting the facts of science, but in habituating the pupil to observe for himself, to reason for himself on what he observes and to check the conclusion at which he arrives by further observation and experiment.

In the late 1870s, the teaching of scientific facts still seemed to be what most educationists wanted for working-class children. HMI Matthew Arnold noted that subjects such as physiology, mechanics and botany had been included in the curriculum for older children. In the elementary school curriculum, because it was generally agreed that 'an entire ignorance of the system of nature' was a 'gross defect' in the children's education, he advocated simple instruction in the facts and laws of nature.

In the nineteenth century there was little consensus of opinion as to what young working-class children in school should be taught and also how young children should be taught. Some educationists placed emphasis on the importance of concrete experiences and exploration whilst others continued to think of early-years education in terms of instruction and facts to be learned. The way in which natural history and simple mechanics were taught would suggest that most teachers and policy makers perceived science education to be concerned with the transmission and learning of facts. This was especially so in the case of working-class children. The development of scientific-process skills had little place in most infant and elementary schools.

The education of girls and boys continued to be similar before they reached the age of 6 years, after this age the curriculum tended to be different. Girls spent far more time engaged in domestic activities, often to the detriment of other areas of learning. This did not go unnoticed but little was done to rectify the situation, possibly because of the opinions current at the time regarding the different capabilities and aptitudes of girls and boys and also about the 'nature' of women and the deleterious effect of learning upon pregnancy. On the more positive side, the work of educationists such as Owen and Froebel had introduced the idea that young children learn best through experimentation and exploration.

Education in the twentieth century

The twentieth century has seen a number of changes in early-years education. In relation to teaching methods there has been a shift in emphasis from the very formal, rote-learning methods which typified the nineteenth century to child-centred, discovery methods. As regards science in the curriculum, this too has changed although not as much as one would expect. For example, until the introduction of the National Curriculum, the natural sciences still tended to predominate in early-years classes.

School sciences as the study of living things

Whilst there may have been changes in the way in which science was taught, the content of science in schools changed little until the 1980s. For much of the twentieth century 'nature' continued to be the main focus of science education in the early years with only occasional forays into the realm of the physical sciences. Until the 1970s the validity of this particular emphasis in science education was not convincingly challenged either by HMIs or other policy makers.

The 1908 Committee stated that student teachers training to teach in nurseries should undergo 'a thorough course of nature study, at any rate of the most common forms of animal and plant life, so that she may stimulate the children's interest and answer their questions intelligently' (Board of Education, 1908b).

Officialdom continued to stress this aspect of science until the late 1960s. Even the Plowden Committee emphasized that 'there is hardly any material more suitable for study by young children than living forms' and were keen to avoid underestimation of the 'importance of the opportunities of natural history' (Plowden Report, 1967). Educational resources and books for early-years teachers offered ample support to those wishing to teach about living things but very little for those who wanted to venture into the field of the physical sciences. Until the late 1960s, in nursery and infant classes, 'science' and the study of living things were thought to be synonymous.

This emphasis on plants and animals meant that for many children their first experiences of the physical sciences was in the upper junior school or at secondary school. The implications of this are discussed in the following section.

Science teaching in the twentieth century

A typical infant class at the turn of the century was still subjected to object lessons which continued to be an exercise of memory rather than an opportunity to explore the natural world. Children in nursery schools were not so restricted but there was a definite emphasis on the physical care of nursery children rather than on their intellectual development.

As in the nineteenth century, the emphasis was different with regard to middle-class children. Children in Froebellian-inspired kindergartens, were encouraged to play and to learn by doing. Most kindergartens, however, were limited to those who could afford the fees.

Rote-leaning and 'cramming' continued in elementary and infant schools for economic reasons rather than because of any commitment to a theory of learning. The Revised Code of 1862 had indirectly inhibited curriculum development and innovations in teaching methods in the early years, and the continuance of grants based on the number of children in each 'standard' led to LEAs pressurizing teachers of young children to 'promote' children as young as 6 because children in the upper school received a larger grant. In this context child-centred methods were bound to be rejected.

After 1908, grants for infants and older children were the same and there was a growing commitment to activity-based education. The development of the 'progressive movement' in relation to early-years education has been well documented elsewhere and will not be traced here but is mentioned because the changing views on children's learning and the role of the teacher influenced the pattern of science education in the early years. The increasing emphasis on activity, exploration and 'free' choice should have meant that all children had the opportunity to engage in a range of activities and explorations which would develop their scientific understanding and the skills associated with science. The reality was somewhat different for a number of reasons.

Firstly, all teachers, albeit with varying levels of awareness, structured the children's school environment. In so doing possible avenues of investigation were opened up whilst others were closed. Young children's explorations were to this extent limited by the teacher. Even within the structured environment, however, children were not 'free' to engage in any activity as the children's peers, parents, the media, etc. all gave children strong messages about what was considered 'appropriate' for them to do. In addition, teachers influenced children's choices in a range of ways (e.g. the way that they presented activities, their expectations of the children's interests and abilities and the support they gave children). Many early-years teachers have recently become aware of the way in which these constraints operate and their educational implications. It is this realization that has led early-years teachers to look closely at children's play preferences, attempt to analyse the factors influencing children's choices and develop strategies to encourage all children to have a wide range of experiences.

The child-centred philosophies influencing early-years education also placed new pressures on the teacher with regard to science. In the nineteenth century and the early decades of this century, the method of teaching science through object lessons did not require the teacher to have any understanding of science, either of the generally accepted 'facts' of science or of the processes by which scientific understandings are developed. Child-centred, activity-based, discovery learning changed this. After all, if children were to be allowed to investigate and explore who knows what questions they might ask? Furthermore, if science was no longer to consist of memorizing facts what were the processes and skills one needed to develop to find out about the world? What, if anything, set scientific exploration apart from other types of exploration?

Teacher education did little to adequately equip teachers for teaching science. In the 1960s, as now, the vast majority of early-years teachers were women and many had given up science at the first available opportunity. Teachers were afraid of science and were afraid to teach it for three main reasons:

1 They were afraid of not being able to answer children's questions
2 They felt unable to help children find the answer
3 They were frightened of the maths that may have been involved (Machin, 1961).

There was also a tendency for women students who had dropped maths as early as possible to opt to teach very young children rather than juniors (Ahmed, 1965). The comments made by Ahmed with respect to attainment in maths and choice of age range have since proved to be equally applicable to the teaching of science in the early years:

> . . . if this tendency is not checked then it will have an extremely undesirable effect on the teaching of mathematics at the Infant and Junior school level. The teaching of maths at these levels is as important, if not more so, as at any other stage in the child's educational career – because attitudes, whether favourable or unfavourable, towards the study of mathematics are formed in the early stage of a child's education, and it is important that at this stage the teacher should know his subject, should have competence in the skill of presenting his subject and should know something of the psychology of learning mathematics.

Books written for teachers did little to suggest that science and technology were areas in which women could excel. One particular book on infant education (Boyce, 1962), written by a former HMI and lecturer, suggested that if a child found something that the teacher did not know about then the child should take the object home in order to 'consult father'! This sort of practice would have given children strong messages about the capabilities of women in science and technology.

Whilst it could be argued that all children suffered because of the lack of good science education in the early years it is possible to argue that the implications for girls were far more serious than for boys. Science has long been seen as a 'masculine' domain, although biology and botany have been defined as 'less masculine' than the physical sciences. From the time that boys were very young it was socially acceptable, even desirable, for boys to engage in the type of scientific activities which were readily identifiable with school science. Small boys, unlike small girls, were also expected to be interested in science and technology. Even if boys lacked experience of science in the early years of their schooling they were encouraged to study science at secondary school whilst girls did not receive the same encouragement. The message that science is 'masculine' did little to motivate girls to excel in it and the ease with which they were allowed to gave up sciences at an early age merely confirmed that science was for boys.

Had girls had a good grounding in science in the early years it is possible that more would have possessed confidence in their own ability as scientists and an interest in science, both of which would have enabled them to deal with social pressures and pressures in the classroom (e.g. boys' domination of teacher attention and apparatus and anxiety about asking questions, etc.). Girls who decided to drop physics and chemistry at the age of 13 and opted to study biology may have done so partly because biology had a less masculine image than physical science and partly because biology was the only science they were likely to have had much experience of in nursery and primary school and therefore was the subject in which they had been able to develop a certain degree of confidence and to which they had developed a positive attitude. Girls were also likely to have been influenced by the fact that as early-years science focused on the biological sciences it resulted in young children having no role models of women engaged in explorations in the fields of physical science.

Present situation

Science and technology currently have a high profile for a number of reasons. Firstly, the government is pressing for science education in schools in order to ensure that Britain does not suffer economically as a result of lagging behind other countries. Secondly, the majority of early-years teachers would state that it is important to be scientifically literate in order to function effectively in the present-day world and to understand the implications of new scientific developments. Thirdly, many feminist teachers are keen to ensure that girls and women are able to play an active part in influencing the direction and nature of scientific and technological developments.

According to research by Tizard *et al.* (1988), it would seem that in reality many children in infant classes spend a very small proportion of their time involved in activities which could be defined as scientific (e.g. testing materials). Furthermore, in many infant schools most science has been orientated towards nature work. A recent HMI report painted a slightly better picture although science education in 66 per cent of infant schools was judged to be inadequate (DES, 1989h).

In nurseries, children are more likely to be engaged in scientific types of activities, but teachers are often not aware of the scientific potential of various toys and activities and as a result there are many lost opportunities for developing children's scientific understanding and skills.

Why have early-years teachers been willing to leave science to luck whilst maintaining that other areas of the curriculum, such as literacy and maths, require high teacher input? One explanation could lie in the different status early-years teachers give to different kinds of knowledge and skills. The three Rs are accorded high status but some early-years teachers are not convinced about the value of science education in the early years. An alternative explanation could lie in the fact that many early-years teachers are women who were 'switched off' to science at a relatively early age. Many such teachers are afraid of teaching science and are unclear about what it is that children should be learning.

Does it matter that early-years teachers do not focus on science? The introduction of the National Curriculum has meant that infant teachers no longer have a choice in the matter but nursery teachers can still choose to make only minimal provision for science. Showell (1979) wrote that it 'is impossible to avoid science with active inquisitive youngsters around you – and many teachers have been teaching science without realizing it'. This may have been an attempt to assure early-years teachers that they are capable of teaching science but it was also the kind of comment that permitted early-years teachers to virtually ignore science. After all, if it is true that all young children behave like scientists in that they observe, ask questions, explore their environment, form hypotheses, etc. then surely all the teacher has to do is ensure that the environment is stimulating and that there are sufficient resources (e.g. magnets, magnifying glasses).

There are a number of reasons for not supporting this virtually non-interventionist approach to science teaching in the early years, four of which are discussed below. Firstly, it is highly questionable whether it is feasible to expect children to 'discover' generally accepted scientific facts virtually unaided. Secondly, this *laissez-faire* approach to science, with the relatively low levels of teacher involvement in scientific explorations, generally means that the involvement of children in 'scientific' activities is not monitored or analysed. As a result some children do not have the range of experiences that the teacher assumes that they have. These differences in involvement have far-reaching implications for equal opportunities in schools, especially if the reasons for lower levels of involvement can be traced to factors such as race, gender or social class. This point is discussed in more detail in Chapter 3. Thirdly, the degree to which a teacher is involved in a particular type of activity has been shown to influence children's involvement and perception of that activity. If female teachers consistently avoid 'scientific' type activities or tend to act in supervisory or servicing capacities girls' involvement in these activities tends to be correspondingly low (Whyte, 1983). It is true that children's perceptions of activities as 'girls only' or 'boys only' is not solely dependent upon the teacher's behaviour, but by avoiding involvement in scientific investigations women teachers are not challenging the view that science is for boys and men. Fourthly, low levels of teacher involvement in science also tends to mean that classroom resources (e.g. books, posters, TV programmes, etc.) are not subjected to the same careful critical analysis afforded to, for example, story books. An obvious result of this is that the image of science, being presented to children, may be predominantly one of White males in white coats working in labs discovering objective, scientific 'truths' or 'facts'. Teachers could begin to change this image if learning resources were more carefully analysed.

The nature of science

The point about the 'objective' and 'masculine' nature of science is one that needs a little further investigation. There has recently been a great deal of discussion about the 'objectivity' and 'masculinity' of science. The link between equal opportunities in the early years and philosophical discussions about the nature of science may not be very clear but it is obvious that a teacher's views about the nature of science will influence her practice. Her perception of science will also be reflected in the strategies she chooses to adopt in an attempt to ensure that the quality of science education experienced by the children in her class is not adversely influenced by factors such as race, gender, social class and language.

Many teachers believe that the content of science is gender-neutral and culture-neutral and it is merely the image of science that alienates girls. These teachers tend to adopt equal opportunities strategies which do not alter the science content but present it in such a way as to take girls' experiences and interests into account (i.e. make science 'girl-friendly'). Whilst this approach may be a valid way of ensuring girls' involvement in science, there is the possibility that this approach makes assumptions about what constitutes girls' interests and perpetuates the view that girls will only be interested in science if it is related to traditionally 'female' areas.

Some teachers believe that science alienates girls because of its emphasis on competition rather than co-operation and because of the way that science is distanced from social and moral issues. These teachers would place more emphasis on co-operative working where individuals' views are welcomed and encouraged. Furthermore, the social, emotional, ethical and moral issues surrounding science would not be ignored and sexism and racism in science would be highlighted and explored. For example, teachers would engage children in discussions about the uses and abuses of science and encourage an understanding of the mechanisms by which science research is sponsored and the effects of this. It is true that it is easier to discuss many of the issues listed with older children but teachers who believe in the subjective nature of science have been successful in raising some of the issues with infants.

A few teachers are beginning to question the 'objectiveness' of science:

> Specific rules of empirical inference and all rules of scientific procedures must prove ambiguous, for they will be interpreted quite differently according to the particular conceptions about the nature of things by which the scientist is guided. . . . For within two different conceptual frameworks the same range of experience takes the shape of different facts and different evidence. (Polyani, 1964)

The implications of such a challenge to the nature of science are far-reaching and have not been explored with respect to early-years science education. It is relevant to note here that the National Curriculum requires teachers in the early years to ensure only that children develop certain scientific concepts and process skills (e.g. observation, hypothesizing, testing, recording). Young children are not to be made aware of the existence of alternative scientific explanations or to be introduced to different views about what constitutes a 'scientific' enquiry. It is difficult to see how such an approach is conducive to equal opportunities as it would seem to necessitate the exclusion of those groups whose views of science differ from that embodied by the National Curriculum.

An admission that science is not neutral but is influenced by social, economic, cultural and political factors would call into question what science we teach, how we are to teach it and why science is taught. Schools, however, do not exist in isolation from society and before real changes in science education occur fundamental shifts in how science is perceived will need to occur in society as a whole. Few members of the scientific community, and even fewer of their sponsors, would want science to be redefined in this way as it would radically undermine the power they currently hold.

CHAPTER 3

Achieving Equality in the Science Education of Early-Years Teachers in Initial Teacher Education

If we are to consider how equality is to be achieved in education we must look at the specific ways in which women, Black and working- class students underachieve in the system. We must also look at how our education system, and perhaps science and technology education in particular, is itself underachieving. This perspective contrasts sharply with the child/student deficit model which has often been used to explain inequality in the past. Here science and technology education becomes the focus of our attention.

Anxiety has been expressed about the economic implications of the underachievement of science education. According to the findings of the International Council of Associations for Science Education (ICASE) Britain's low standing near the bottom of the international league table for science has resulted from the inadequacy of science provision in our primary schools. ICASE research showed that less than half of all primary teachers had studied science beyond the age of 13 and less than one in ten of those teaching 10-year-olds had science as their main subject in teacher education (*Times Educational Supplement*, 1989). While the following arguments focus upon science and science courses, many of the points made are equally relevant to mathematics education and to elements of design and technology courses.

It is predominantly women who teach in early-years education.

The vast majority of students in initial teacher education (ITE) establishments who opt for infant and nursery teaching continue to be almost entirely women. In the light of the substantial research evidence on the underachievement of science and technology education we must accept that many of our student teachers on early-years courses are not confident with science and technology and may even hold negative views about their capacity to learn these disciplines, let alone to teach them! This does of course have serious implications for course design and delivery on the part of science and technology departments in ITE. Student teachers need to understand why they may feel inadequate in some subject areas in order for them to prevent the same occurring to their pupils. They also need to share experiences with those students who feel more confident with science and technology. Research evidence needs to be used to promote student teachers' understanding of their own achievement or underachievement.

The Swann Report (1985) argued that the underachievement of Black children was largely due to racial discrimination, particularly in employment and housing (2.3) and partly in education itself (2.4). The Swann Report made it clear that, if we are to tackle this underachievement we must accept that we are faced with a dual problem. Firstly to eradicate White racism and secondly to provide equality of opportunity in education (2.6). These conclusions are equally applicable to both gender inequality and the underachievement of working-class children.

Before women were given the right to vote in Britain on the same basis as men, the status quo was defended by citing pseudo-scientific (and now rejected) eugenic arguments that women were innately inferior. The very same arguments were used in the nineteenth century against extending the franchise to the working classes and in more recent times similar pseudo-scientific racist views have been popularized by theorists such as Hans Eysenek and Arthur Jensen. These prejudiced ideas live on in popular culture. 'Race',[1], class and gender inequality is still rife in our society. If we are to provide genuine equal opportunities then some radical changes in the education of our children are needed. Student teachers need to understand the dynamics of inequality and appreciate that young children need to be socialized within an environment characterized by positive images and approaches to 'race', gender and class.

Traditional compensatory perspectives stressed the 'deprived' conditions in which some children were socialized and sought to provide extra nursery places or English language provision. The solution was seen in terms of greater quantity or quality in education, rather than questioning the educational context itself. We must begin to recognize that a child's first 'teacher' is the parent and it is therefore essential to involve the parents in counteracting inequality. Teachers need to work with parents and the community to ensure that more of the child's experience is shared. A curriculum emphasizing equality is not enough on its own. If we are to 'achieve' we need to provide an all-embracing equalitarian ethos within which our children are educated. The challenge to the teachers is thus, not only to formulate strategies to make science and scientific products accessible and relevant to children's lives and thereby a part of their everyday language, but also to accept an educative role with the parents.

In our society sexist and racist structures, images and practices pervade the experiences of our children not least in the context of science and technology. Adult roles and the effects of media images are especially pervasive, up to a quarter or more of our children's waking hours in the infant years is often spent in front of a television screen. Young children learn through role models, by doing, talking, and applying overt and subliminal messages. For example, when 'Mum' has trouble with a domestic appliance it is often time to call in 'the man'. When children leave their parents and begin nursery or school at the age of 3 or 4 they are, in many significant senses, the product of their individual and unique previous experiences. New knowledge and understandings must grow out of these previous experiences, the ongoing out-of-school experiences and the new experiences that we provide in school. Piaget stated that children need to constantly 'equilibriate', they need to match their understanding with their observations of the environment around them. Margaret Donaldson adds that this needs to be within the social context of the child.

The curriculum must address itself and relate directly to inequality in the home and out-of-school environment. This can be achieved to some degree with the use of illustrations, stories and activities engaging children in alternative roles and practices. Children need a variety of scientific experiences if they are to gain confidence. These experiences will clearly be much more successful if they involve the parents as well as the teachers.

A study of one school by Alan Peacock in 1989 could probably be replicated across the country. The study suggested that parents offer very mixed responses to questions such as 'What is science?', 'How is science taught?' and 'When should science start?'. Peacock emphasized that parents wanted to know much more about what went on in the classroom. He noted that:

> there were noticeable differences between parents of younger and older children. Those with the very young were sure they didn't do any science during topic work, whilst those with the older children thought they did. (Peacock, 1989)

This was because parents' perceptions of science were based on their own school experience of the distinct subject areas of biology, chemistry and physics. In fact the school was involving all its pupils in science activities. Clearly the teachers were failing to involve and communicate adequately with their parents.

Within ITE science courses, questions and issues of inequality are neglected, its effects upon content and teaching and learning processes is ignored. As Michael Day put it in a 1989 Commission for Racial Equality occasional paper:

> What is apparent is that compared with schools, universities and polytechnics have been relatively untouched by the debate on racial equality in education and have not, on the whole, seen the need to develop specific policies in this area. It may be that these institutions have seen themselves as incapable of discrimination or unequal treatment, and thus absolved from discussions of inequality in access to educational opportunity. (Williams *et al.*, 1989)

The idea that science education involves passing on a body of knowledge which is value free has been the overwhelming and prevailing ideology of science educationists. In ITE science courses, any consideration of the social factors which influence learning tend to be considered the realm of education and teaching studies courses alone. Student teachers are rarely introduced to any research which raises the issues of inequality with regard to 'race', gender or class within their science or technology courses and are even less likely to discuss course content and the effects of teacher attitudes in this context.

Recently a great deal of concern has been expressed about science teaching in the primary school. Particular attention has been directed to the lack of time spent by teachers teaching science, the poor quality of their scientific knowledge and the lack of confidence felt by teachers (Nash, 1989). If teacher educators are to take the issue of equality into account in initial training then science courses must consider issues of equality and community contexts as well as teaching strategies and learning processes. An attempt is made in the following paragraphs to define these requirements further.

Equality

Student teachers should be aware of the fact that science and technology are not exclusively a White, European and male phenomena. Tutors should offer appropriate exemplars. As the National Curriculum Non-Statutory Guidance states:

> People from all cultures are involved in scientific enterprise. The curriculum should reflect the contributions from different cultures, for example, the origins and growth of chemistry from ancient Egypt, Greece and Arabia to the later Byzantine and European cultures, and parallel developments in China and India. It is important that science books and other learning material should include examples of people from ethnic minority groups working alongside others and achieving success in scientific work. Pupils should come to realise the international nature of science and the potential it has for helping to overcome racial prejudice. (Sect. A10: 7.8)

It is of course equally important to ensure that the achievements and contributions of women scientists are also included. Student teachers should be helped to identify the popular ideologies of science and technology, one important example being the mistaken conception of a linear continuum between simple and sophisticated scientific and technological societies. Student teachers should also discuss the dangers of adopting a simplistic conception of social 'progress'. While written in the context of 'Ethnic Minorities', the following statement from the report of the Design and Technology Working Group (DES, 1989d), could also apply equally well to gender:

> The variety of cultural backgrounds of pupils can broaden the insight they all have into the range of appropriate, alternative solutions to perceived problems. There are rich opportunities here

to demonstrate that no one culture has a monopoly of achievements in design and technology. (Sect. 1.46)

As the document goes on to clarify, such issues are just as relevant to those schools where there are few Black pupils and there is an urgent need to broaden the curriculum in many single-sex boys' schools.

Community context

Student teachers need an understanding of the cultural and linguistic dissonance of their pupils and they will need help in identifying strategies for making science accessible to all children. They must understand the role of parents as the child's first teacher/s and the ways in which the attitudes that children bring with them to any science or technology work effects their perceptions. Given the sexist and racist role models which children assimilate, student teachers need to manage group work carefully to maximize the breadth and continuity of each child's scientific and technological experience. If *all* of our pupils are to achieve, then student teachers must consider carefully their use of encouragement, language, teaching materials and peer grouping practices. Many White middle-class student teachers will also need to gain an appreciation and knowledge of the contemporary community context within which our children live.

Teaching strategies

Early-years student teachers need to have a good knowledge of the National Curriculum and Non- Statutory Guidance for science. They will also need to develop skills and strategies for working collaboratively with other colleagues, sharing strengths and planning curricula as the National Curriculum encourages. These initiatives also demand practice and confidence in integrating science with topic work.

Learning processes

Student teachers should have an understanding of child development and be able to apply this to the organization of the science and technology curriculum. They need to understand the role of learning theories and how these should effect the delivery of content and processes of assessment. They must be taught the importance of experiential and active learning and the value and significance of cross-curricula approaches. They must also be aware that language acquisition and development is a fundamental process to any learning.

Unfortunately the reality of many early-years science courses is far from these ideals. In fact it could be argued that many science courses contribute to the perpetuation of inequalities and the lack of confidence and underachievement which many early-years teachers feel and undoubtedly pass on to the children they teach.

Science courses have traditionally failed to address either community contexts or equality. The main part of science courses are concerned with skills and knowledge. These have traditionally been delivered by largely White, middle-class and male-dominated departments of science, most of the tutors having only had experience in secondary education and some in junior schools. The hidden messages received by early-years student teachers on the whole are that science is a male-dominated area (which immediately excludes them as scientists), and that any early-years science appears to be 'watered down' junior science. They are often given no specific guidance on early-years science at all.

Many science departments continue to retain the artificial boundaries between sciences often referring to them as 'life' science, 'physical' science and 'earth' science. This effectively reminds student teachers of their own secondary school days and how they felt about science then. Research shows that women teachers prefer to teach nature-based science, and as a larger proportion of girls take biology at GCSE and A Level, this suggests, unsurprisingly, that teachers prefer to teach those areas they feel most confident in.

Since the formation of the Committee for the Accreditation of Teacher Education (CATE), Bachelor of Education (B.Ed.) courses in particular have been forced to change. The emphasis on student teachers having to study their chosen main subject to a high academic and specialist level has resulted in student teachers spending 50 per cent of their course time on their specialism. In many cases this has more than doubled the time previously spent on specialist study. This has meant a cut in other course areas. Curriculum studies which include the science courses have been cut marginally but the most drastic cut in hours has been made to those courses which have traditionally dealt with the social factors which influence children's learning. The two main areas being education and teaching/professional studies (hours have been more than halved in many cases) and some school-based work. In most institutions the one-year Post Graduate Certificate in Education (PGCE) course follows a similar course content to the B.Ed., yet even less time is available to cover these issues.

Many B.Ed. and PGCE courses only offer their students between 30 and 60 hours of science education. Many of these science courses do not include design and technology. As a result of the introduction of the National Curriculum CATE has issued new criteria for courses and these increase the hours ITE student teachers spend on science, design and technology to bring them in line with the other core subjects of mathematics and English. On the initial B.Ed. this means 100 hours of college-based work will be devoted to science, design and technology. Such courses may often be called 'Studies in the Environment' and may deal with other subjects such as history and/or geography thus reducing the overall time spent on science, design and technology. This kind of cross-curricula approach is of course desirable as long as it is not achieved at the cost of a proper preparation for teaching science, design and technology. PGCE students will also spend 100 hours of their time on science, design and technology. Unfortunately it seems likely that these new CATE proposals will lead to the educational studies provision on PGCE courses suffering even further in the future.

Curriculum science courses generally focus on process skills such as classification, observation, prediction, measurement and experimentation. These are usually drawn out through the study of particular themes such as ourselves, air and water, weather, toys and growth. Quite often early-years student teachers follow the same course as the middle-years student teachers for the first year or two before age specialisms are chosen. It is, however, clear that issues of 'race', gender and class are not seen as relevant areas of study in relation to science teaching and learning. Even where themes are followed which may appear to beg an anti-racist or anti-sexist approach, such as 'ourselves' or the 'cycle of life', opportunities are missed because of the way science tutors interpret science or because of their lack of knowledge and skills to deal with the issues appropriately.

ITE institutions have begun to rise to the challenge of the National Curriculum and cross-curricula themes but are failing to include issues of 'race', gender and class. It is apparent that ITE has so far failed to recognize the central issues of inequality, let alone deal with them effectively. In a 1988 Occasional Paper, the Anti-Racist Teacher Education Network (ARTEN) discussed the viability of 'permeation' as a model for change in ITE institutions. The document states:

> Consensus in our society is achieved through the suppression of struggles based on race, class and gender. The institutional reality for those concerned about promoting genuine antiracist perspectives in schools and colleges is that permeation as a model for change cannot work whilst it is being 'implemented' by people who have not raised their own consciousness and understanding of issues of race and racism. (p. 4)

In terms of producing early-years teachers who are also confident and successful science practitioners the ITE institutions have again failed. Despite the regular indictment of evidence published in educational papers and journals showing how little science our colleagues teach, and how low their confidence is, the situation has not changed since the 1978 HMI survey on primary schools. The situation has not altered significantly in the last decade.

ITE institutions have been repeatedly criticized by HMIs for not recruiting more early-years specialists. The relevance of this is apparent to most student teachers undertaking early-years courses and especially to those taking science. An effort to recruit more early-years specialist tutors could also increase the number of women who are grossly underrepresented in ITE. It is important that institutions do not depend solely on the criteria of higher degrees to recruit staff. This discriminates against women applicants who, while as capable as their male counterparts, have not, for various reasons such as a break in career to raise a family, had the same time or opportunities to study for and gain higher degrees or publish papers and articles. Black people are also underrepresented in ITE and positive action needs to be taken by institutions to rectify this situation. It should be made clear that the opportunity to follow a higher degree would be given during employment.

In the present climate, with the demands of the National Curriculum, and science representing one-eighth of the curriculum, the situation has reached a critical point. ITE institutions urgently need to re-evaluate their course work with regard to the content and approaches to science. In addition they ought to re-think who it is that is teaching the subject. Given the limitations imposed by the CATE criteria on education studies courses, cross-curricula provision for equal opportunities and anti-racism would be beneficial. ITE science departments need to look again at what science education means for the 1990s. What it means for student teachers and the children they will teach. This must be considered in the light of a rapidly changing society, both national and international, and the changing role of women and Black people.

As far as the primary years are concerned all children will be required by law to follow the National Curriculum and have a similar set of experiences in science, design and technology. In the secondary years some pupils will spend 20 per cent of their curriculum time on science while others will opt for only 12.5 per cent. Girls are likely to take up the majority of places on these more limited courses if teachers in the early and primary years continue to lack the appropriate skills and understanding to involve girls as much as boys in their compulsory early science experiences.

ITE needs to offer courses which enable student teachers to examine attitudes, teaching strategies and methods. Student teachers must be taught the importance of sharing the child's community. Too many inner city teachers do not live locally and never really perceive the needs of the community group. They impose their own values and perspectives in education, hopefully with the application of the Education Refom Act this will be reduced. Teachers need to break down the barriers of professionalism to meet the needs of their pupils and parents. We need to teach our student teachers diagnostic approaches rather than prescriptive ones.

Moreover, schools and nurseries need to be open in such a way that they can support parents by taking a 'therapeutic' role where appropriate as well as being educative in a non-threatening manner (Poulton and James, 1975). Schools and nurseries often benefit from the provision of a parents' room where teachers and parents, usually mothers, can talk, drink coffee, make educational games, plan activities, and more importantly discuss why such activities involve so much language use, why girls, Black and working-class children need strong role models, why we must all avoid sexist and racist language and practices.[2] Above all, if we are to be effective in countering inequalities in education then student teachers must be taught to understand that this is less to do with what we teach and much more to do with how and where.

A Commission for Racial Equality survey on equal opportunities policies in 68 universities and polytechnics (Williams *et al.*, 1989) found many institutions to have a 'tone of moral superiority or complacency plus ignorance of the issues'. In my view, all ITE science departments should be concerned about issues of inequality and recognize that early-years teachers do not feel confident about teaching science. ITE science departments should give careful consideration to some of the points outlined below. Many institutions will recognize that they are making beginnings, some may have make significant progress in these directions already. This does not call for complacency given the importance of the issues raised in this chapter.

What should science, design and technology courses for early-years students offer? Consider the following ten recommendations:

1 Early-years student teachers should be given a sound grasp of science, design and technology in the National Curriculum set within a critique of the statutory orders. Courses will need to cover at least key stages 1 and 2 if continuity is to be maintained across the infant/primary years.

2 Courses in science should include wider educational issues of how children learn and the impact of various teaching methods. This could be achieved by liaising with the work of the education and teaching courses and the full use of the nursery and infant classroom both within and outside the institution.

3 Colleagues across departments need to illustrate 'girl-friendly' and anti-racist approaches to teaching with special emphasis on subjects not normally associated with inequality such as science. The effects on all children of not taking such an approach should be made clear with the use of research evidence.

4 Recognition should be made of the fact that most early-years student teachers are women and that very few of them will have had experience of physics, design, technology and

chemistry to GCSE level. Science courses may have to be extended in order to bridge this gap.

5 Institutions of ITE need to make a concerted effort to recruit more women and Black staff on to their science team given the importance of role models and the need to counteract stereotypes.

6 Certain modules within early-years science courses should deal with equality issues and science, stressing the community context. Permeation has been shown not to work (ARTEN, 1988).

7 Modules which recognize the above need to be created. They must be accessible and flexible enough to cater for a variety of training needs, that is, not just B.Ed. and PGCE, but for school-based in-service and the education of licensed teachers.

8 ITE departments must engage in staff development in this area and visit other departments to observe and discuss good practice. Many more tutors should take advantage of the 'recent and relevant' school experience by working in multi-ethnic, urban schools and teaching, where possible, infant or nursery classes.

9 Science departments should try to use materials and resources with equality perspectives. They should also be active in putting pressure on publishers to produce such materials.

10 Science courses need to emphasize the framework within which the National Curriculum will be taught, including the cross-curricula issues of multicultural education, gender, special educational needs and the basic concept of 'entitlement' to education by each and every child to a broad and balanced curriculum.

Notes

1 The term 'race' is given in inverted commas to indicate the problematic nature of the concept. The notion that distinct 'races' exist has been scientifically discredited as:
   (i) the extent of genetic variation within a population has been found to be usually greater than the average differences between populations;
   (ii) while the frequency of occurrence of different alleles (forms taken by genes) varies from one 'race'; and;

(iii) inter-breeding and largescale migration has caused the distinction between 'races' identified in terms of polymorphic (dominant gene) frequencies to be blurred. See Robert Miles, *Racism and Migrant Labour*, p. 16. RKP (1982).

2 Here Bernstein can be usefully read to clarify the way that the classifications and insulations between 'family', 'socio-economic context' and 'education' need to be weakened. See Basil Bernstein, 'Codes, modalities and the process of cultural reproduction: a model'. *Language and Society*, 10, 327–63. *Class, Codes and Control*, Vols 1–3, 1971–73, RKP (1981).

CHAPTER 4

'Girls' Stuff, Boys' Stuff': Young Children Talking and Playing

NAIMA BROWNE and CAROL ROSS

Background

This chapter is based on conversations with, and observations of, a large number of very young children in nursery and infant classes. It explores how young girls operate within nursery and infant schools and the degree to which gender influences children's involvement in science-orientated activities. Many of the observations relate to constructional play, as this is an area with a high science, maths and technology content. Furthermore, it is an area of concern for many teachers as they believe that the activity is dominated by boys.

The nursery and infant classes differed in terms of their cultural, linguistic and ethnic composition. Most classes had approximately equal numbers of girls and boys, but in a few classes either girls or boys predominated. The adults in the classes varied in respect of their awareness of, and commitment to, the various issues connected with equal opportunities in education. The observations took place over a period of four years and a number of issues emerged which are particularly pertinent to those concerned about the science education of girls in the early years of schooling.

From a very young age children seem to have clear ideas about what girls do and what boys do

In our observations of the play preferences of nursery and infant children there were clear demarcations along gender line. Figures 4.1 and 4.2 provide an indication of how marked this demarcation was in the case of two activities, creative activities (e.g. painting, collage, printing, etc.) and constructional play. The various patterns of children's play preferences have been discussed elsewhere (Clarricoates, 1980; Walkerdine, 1989; Whyte, 1983; Thomas, 1986) and need not be discussed here.

Our conversations with children in nursery classes suggested that children as young as 3 were very conscious of what they

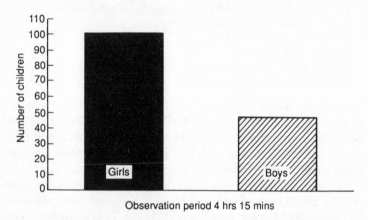

Observation period 4 hrs 15 mins

Figure 4.1 Involvement of girls and boys in creative activities

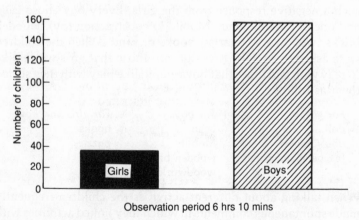

Observation period 6 hrs 10 mins

Figure 4.2 Involvement of girls and boys in construction play

played with and appeared to have mapped out in their mind which toys and activities were 'for girls', which 'for boys' and which were gender-neutral.

Many of the infant children also held very strong views. The following exchange was typical of many conversations with 6-year-olds.

Boy 1: I don't play with dolls – they're girls' stuff.
Adult: What do you mean 'girls' stuff'?
Boy 2: It's girls' toys.
Adult: Do you think there is anything else that's 'girls' stuff'?
Boy 2: Barbies and Sindies and painting is girls' toys.
Girl 1: Some boys can do painting.
Adult: What's 'boys' stuff'?
Girl 2: Action Force.
Boy 1: Yeah, and transformers.
Girl 2: Ludo's for girls and boys.

Some of the infant and nursery children were shown a selection of photos of different toys and asked to choose two toys with which they would most like to play. In an effort to avoid influencing the children, the toys had been photographed against a plain background without packaging and no child was shown in the picture. Despite these efforts the resultant choices were frighteningly predictable. Every girl chose a doll and then one of the following: felt pens and paper, a doll's house, jigsaw puzzles, books or sand. One 6-year-old girl's immediate response on seeing the photo of the Lego was to say, 'I *hate* it'. No other toy elicited such a negative response from the girls. Every boy chose Lego and one of the following: Mobilo (a construction toy), wooden bricks, woodwork, puzzles, books or sand. When the children were asked to sort the toys out into those that girls would like to play with and those that boys would like play with they sorted them as follows:

For girls	*For boys*	*For girls and boys*
doll	Lego	books
doll's house	Mobilo	puzzles
felt pens and paper	wooden bricks	sand
	woodwork	

When talking about different activities the children frequently made spontaneous comments in which they linked activities with gender. This would seem to indicate that very young children understand that gender is used as a means of organizing people and society.

Identification of activities in gender-related ways affects how children use materials

On observing girls engaged in constructional play we began to suspect that girls' and boys' learning experiences may be quite different even when they are using the same materials. It was apparent throughout our observations that not only did young children have clear ideas about what girls and boys play with, but they also had firm ideas about how resources were to be used by each gender.

In one infant class we observed a group of girls and boys as they settled down to play with Lego. There was a total assumption on the boys' part that the girls would make houses and the boys would make vehicles or guns. This was clearly demonstrated by the way in which the boys immediately, and without discussion, sorted out the Lego, giving the windows, doors, base boards and larger bricks to the girls and the wheels and rotatable connectors, etc. to the boys. The girls did not protest, although one girl who had planned to make something other than a house changed her mind saying, 'Oh, all right, I'll make a house'.

When children in another infant class were shown a photo of Lego the comments they made revealed that they were also very conscious of the different ways girls and boys used it.

Girl 1: I make houses out of Lego . . .
Boy 1: We always make cars.
Boy 2: I make guns too. We make guns and have fights.
Girl 1: The boys in our school always make cars.

Nursery-aged girls and boys also frequently used the same material in quite different ways. The differences in use were related to the children's sense of what was appropriate for each gender. For example, five nursery children were working at a table set up for cutting, sticking and stapling paper. The boys were making loops out of paper which they slipped on to their arms. The girls were cutting out oval-shaped pieces of paper which they stuck on to sheets of paper.

Adult (to boys): Are you making arm-bands?

Boy 1: No, they're muscles.
Girl 1: We're making flowers.
Boy 2: We're making muscles to play He-Man.
Girl 1: They always make muscles.
Girl 2: Sometimes they make He-Man belts.
Adult (to girls): Do you sometimes make muscles and He-Man belts too?
Girl 3: (laughing) That's silly!

The children's view of how certain resources should be used in gender-related ways often effectively excluded girls from particular construction materials which were not seen as having the potential for 'girls' play'. The materials which children identified as being used to make cars, aeroplanes and guns were almost constantly boy-dominated. In the classes we observed, boys tended to use constructional materials in more sophisticated ways, making more use of it as a medium and exploited its three-dimensional properties (e.g. its capacity for movement, balance, its potential for complexity of configuration). The boys also tended to approach constructional play as an individualized activity. Girls, on the other hand, frequently made very simple structures and, moreover, they often used it as a foil for social play. In one class the girls worked co-operatively to construct a simple house and a few basic tables and chairs from Lego. They then fetched some 'Play People' and used their Lego house to engage in play centred around families at home. In contrast, the boys worked on their own and constructed cars, aeroplanes and guns. These Lego constructions were quite complex and many incorporated movable parts. They played briefly with these models using appropriate sound effects and then changed their construction into something else. The boys were engaged in a constant process of making and remaking Lego constructions.

Similar patterns were observed in many early-years classes. This would suggest that girls and boys may focus on two very different processes when playing with construction toys. Girls appear to be more concerned with the process of social interaction and boys with the process of making things. These observations would strongly suggest that ensuring girls and boys engage in the whole spectrum of activities on offer in nursery and infant classes is not in itself necessarily going to ensure that girls' and boys' learning experiences in school are not qualitatively different from each other.

The vicious circle

Children's choices of activity were not solely influenced by judgements about whether or not it was 'appropriate' in gender terms. This may have been an initial determining factor on first encountering the activity but the girls' comments suggested that lack of confidence and feelings of inadequacy about their ability to handle specific activities rapidly became an additional factor which inhibited their involvement.

> *Adult*: Do you play with the Lego?
> *Girl* (aged six): Sometimes . . . I make cars – I can't do it. I make grass, flowers . . . I can make a house. Sometimes I can't do it.
> *Adult*: What do you do then, who helps you?
> *Girl*:   When I can't do it I just go away and do a drawing.
> *Adult* (to boys): What about you, who helps you when you can't do something with the Lego?
> *Boy 1*: I can do it.
> *Boy 2*: I can always do it.
> *Adult*: But there must be times when you find it difficult – you can't make what you want to. What do you do then?
> *Boy 1*: I just do it.

Many of the girls described constructional activities as 'boring' but this may have been their way of opting out as the same girls became quite enthusiastic when given time, space and encouragement.

An analysis of nursery children's first choice of activity was revealing. On entering the classroom at the beginning of the day, many nursery children tend to choose to become involved in those activities they feel most confident about. Many girls chose to play in the 'home-corner', do a drawing, become involved in a creative activity (especially if an adult was present), read a book or talk to an adult. It was extremely rare to observe a girl choose to play with a construction toy as soon as she entered the class whereas it was very common to observe boys doing so.

Confidence and power in relation to gender domains

The notion of territories or domains was highlighted by the way in which the children's confidence and assertiveness varied depending upon whether they were in or out of 'their' territory (see also Walden and Walkerdine, 1982). The degree of confidence was reflected in the way that children approached activities.

> *Boy*: I've made a dragon (shows model made from Mobilo).
> *Girl*: Can you make me one?
> *Boy*: (Incredulously) Can't you do it?
> *Girl*: Do I need this?
> *Boy*: No, you don't need that. You need a yellow one, one of these.
> *Girl*: (Collects pieces to make model) That goes there. Did you put one of these in?

This constant questioning and checking continued until the girl had completed her dragon.

The children wielded power over the other sex when in their 'own' territory. Sometimes this power was apparent in the way the children dictated the terms or the rules of the game. Sometimes it was used to exclude would-be interlopers. Variations on the following exchange were frequently heard in nursery classes.

> *Girl 1* (to two boys playing with large bricks): Can I play?
> *Boy*:   No . . . Well, don't put your hands on the bricks 'cos I've wiped them.
> *Girl 2*: Can I play?
> *Boy*:   No, not *another* girl.

A third boy joined the group and was immediately included in the building activity. When the girl made a suggestion about the structure being built, one of the boys pushed her away and shouted, 'No! I told you not to touch the bricks'.

In those schools where the adults expressed disapproval about exclusion on the grounds of gender, children would offer an alternative justification for the exclusion. In one such nursery school ten boys were playing with large bricks when a girl attempted to join them.

> *Boy*:   This is a rocket, not she stand on it!
> *Adult*: She can stand on it.
> *Boy*:   Who said you could play? Who said you could play?
> *Adult*: She can play.
> *Boy*:   No she can't.
> *Adult*: Why?
> *Boy*:   'Cos she's got a horrible nose.

When in 'their' domain the girls were equally assertive. Three girls were playing in the home corner and a boy appeared at the door of the home corner.

Girl: (Before the boy had said anything) No!
Boy: I want my baby.
Girl: No! Go away.
Boy: (Pleading) I want it.
Girl: (In very adult tones) Here it is. Now run away.

In some classes the adults provided support and encouragement to children who wanted to move into previously unexplored 'territories'. It was interesting to note how children playing in areas they did not feel entitled to, frequently showed anxiety when children who normally dominated the activity began to join in. This anxiety led children to turn to an adult and make requests such as 'Don't let him play', or, in relation to her own model or the construction pieces, 'Don't let him take it'. When children were operating beyond their perceived gender domain their confidence was very easily undermined. For example, a 4-year-old girl was in the process of grappling with a problem with a construction toy when a boy approached. He watched her for a moment and then said, 'That's not how you make car. I know how'. He then began to tell her what to do. The girl responded by adopting a helpless demeanour and made no comment. Following the boy's intervention the adult tried to encourage her to continue but she refused saying 'I can't do it'.

Girls' strategies for gaining access to 'boys' domain

It was encouraging to notice that many girls did not passively accept the gender-based boundaries and developed their own strategies to gain access to 'boys' activities.

Boys who wanted to play in areas usually dominated by girls often ensured that they were able to do so by employing tactics which included disruption and fighting. Girls rarely employed such overt or violent means of gaining access. Many girls would try and gain access by attempting to show the boys that she was 'one of them' or by making it explicit at the outset that she was not going to attempt to dominate the activity and that she was willing to become involved on the boys' terms. In the following example a girl used both of these strategies, with varying degrees of success.

Three boys were playing with bricks. A girl had tried, unsuccessfully, to verbally negotiate an entrance to the activity. She then went away, built a machine-gun out of Duplo and returned to show it to the boys and said, 'I made gun . . . Look! I've got a gun'. The boys, who frequently engaged in gun play, could not resist a quick glance but they made no further response and continued to play with the bricks. The girl returned the Duplo to its tray and came back.

Girl: Can I be the baby girl? (repeated three times)
Boy: No, you can't be anything.

A few minutes passed and the girl remained where she was.

Boy: Oh, all right then.
Girl: Baby girl?

The boy nodded and the girl joined the group.

Gaining access to an activity was often not the end of the battle as those who regarded the activity as 'theirs' often insisted on taking the lead and dictating the terms of play. Sometimes girls were determined not only to play with toys normally dominated by boys but also to play with the toys on their terms. A 3-year-old in a nursery decided that she wanted to play with the trains which, in this particular class, the boys had claimed as 'theirs'. As soon as she started playing she came into conflict with the boys who tried to insist that her train move in the opposite direction. One boy verbally threatened her, glared at her and repeatedly said 'No!'. When this failed to deter the girl he resorted to appealing to a nearby adult for help by crying, 'She's breaking up my train'.

The two examples show the degree of confidence, resolve and tenacity required on the part of girls who want to engage in activities normally dominated by boys.

Gender or culture as the basis for play choices?

In the schools observed, there was no evidence of girls or boys choosing activities on the basis of their cultural, linguistic or religious background. Rather it was the case that children who were presented with a range of new activities tended to be drawn to those they were familiar with until they had developed the confidence to try something new. The pattern of girls playing with dolls and boys with cars and Lego was consistent across cultural and social groups. Gender appeared to be the major factor influencing children's play preferences. Children were also influenced by the degree of language that was an intrinsic part of the activity. If the activity required that the children talk to each other (e.g. solving a problem in pairs) there was a tendency for children from the same linguistic group to play together.

Sabotaging of teachers' attempts to involve girls in constructional play

We found that the girls were very adept at sabotaging teachers' efforts to ensure that they engaged in an activity they would not normally choose. One of the strategies girls employed was to move to a completely different area of the classroom or nursery where the teacher could not see them in the hope, often fulfilled, that the teacher would forget about what she had asked the child to do. Another involved merely sitting near the activity for a short while and keeping a low profile until they judged it was safe to move away. If the task involved constructing something, a third strategy was to ask another child to make the model which they then showed to the teacher. This strategy was not used very often as it required the co-operation of other children. The fourth strategy, which was often used by girls when asked to make something using a construction toy, was for the girl to produce a model of minimal complexity which involved very little effort, problem-solving or thought. Girls were observed to connect a few bricks together and on showing it to an adult often received a great deal of praise. This last strategy was employed frequently and would suggest that at a very young age the girls concerned had worked out that not much was expected from them in the field of constructional activities and that a minimal effort would result in praise. Ironically the teachers who were most liable to be overgenerous with their praise were those who were keen to develop girls' constructional skills.

The 'magnetic' adult

Most cross-domain play occurred when an adult was involved in the activity. Figures 4.3 and 4.4 show the effect of an adult presence on the involvement of girls and boys in creative activities

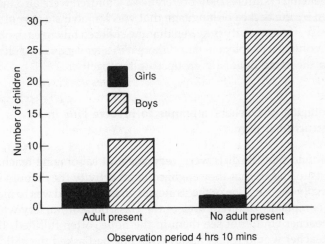

Observation period 4 hrs 10 mins

Figure 4.3 Effect of adult presence on involvement of girls and boys in creative activities

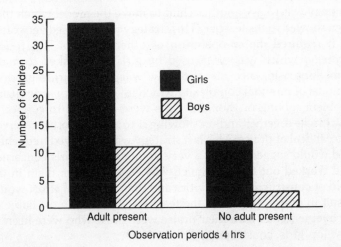

Observation periods 4 hrs

Figure 4.4 Effect of adult presence on involvement of girls and boys in construction play

and constructional play. This pattern has also been observed by others (e.g. Whyte, 1983). The degree to which the adult was a 'magnetic' influence depended upon the extent and nature of the adult's involvement. Adults who were enthusiastic and actively involved attracted more children than did those adults who 'serviced' activities.

In multilingual classes, bilingual children were attracted by adults who spoke their home languages. The children may have been attracted because they welcomed the opportunity to engage in stimulating discussion or because the adult was able to make the activity appear interesting. In nursery classes where there were women and men on the staff, girls were more likely to become involved in a 'boys' activity if a female rather than a male member of staff was present. Similarly, boys were more likely to engage in 'girls' activities if a male member of staff was present. In one class when the male teacher was working with children in the creative area there was a predominance of boys. On other days, when the female teacher was working in the creative area girls predominated. When no adult was present the area was dominated by girls.

Perceptions of scientists and technology

It has been argued that girls 'switch off' from science and technology because both are associated with men and 'maleness'. Whilst none of the nursery children we talked to had formed ideas about what scientists are and what they do, comments made by under-5s suggested that many very young children have begun to develop clear ideas about the scientific and technological abilities of women.

> *Girl* (aged 4): My sister got a typewriter.
> *Adult*: Do you play with it?
> *Girl*: Mmmm. . . . It's broken.
> *Adult*: Oh dear, how did it break? Can you fix it?
> *Girl*: (Laughs) No!
> *Adult*: Why not?
> *Girl*: 'Cos I can't.
> *Adult*: What about your Mummy, could she fix it?
> *Girl*: No! My Daddy does.
> *Adult*: Why not your Mummy?
> *Girl*: 'Cos she's no good.

Not many of the infant children had clear ideas about what scientists do. The following is a record of a fairly typical conversation with a group of 6- and 7-year-olds.

Adult: What is a scientist?
Boy 1: A scientist? I dunno.
Boy 2: It's a man . . .
Boy 3: It's a man, it's a doctor and he helps women and helps children, he gets his case and helps them.
Girl 1: It's a doctor, a woman (whispered), no, a man (said loudly).
Girl 2: It's somebody who finds out what makes people sick. Testing things like yoghurt because hazelnut yoghurt got a poison in it and makes people sick. (This was a reference to a recent outbreak of botulism which was traced to hazelnut yoghurt.) He tests water. It's a man. It's a man who tests medicines and make-up. He works in a school laboratory.
Girl 3: You need science for doctors and nurses.
Girl 4: It's like Doctor Who. They go in space.

Most young children assumed that scientists were male and explained this by saying, 'Because they are'. It seems clear that for many children, science and science-related activities are perceived as a male domain. However, this perception is influenced by the types of activities which are identified as being 'scientific'.

Concept development and gender domains

The fact that science is perceived by many children to be about men in white coats doing experiments and going into space may result in girls feeling unable to engage in overtly scientific pursuits. However, activities defined as being 'for girls' also involve a range of scientific concepts. Our observations indicate that girls and boys often apply similar conceptual understandings in different situations. Many apparent gender differences in scientific under-standing may have more to do with the context of conceptual application than with differences in conceptual development. This consideration assumes a new importance with the introduction of the National Curriculum and the 'testing' of very young children in order to ascertain their levels of attainment.

Girls engaged in a range of activities associated with girls and women (e.g. cooking, sewing, some forms of craftwork) are introduced to a wide range of scientific concepts. It is clear from talking to young children who have had experience of cooking that they are developing scientific understandings related to the properties of materials (e.g. their appearance, the effect on them of different temperatures, how they combine with other substances). Similarly, it was apparent from watching a girl making a dragon from junk materials that she had learnt a great deal about the characteristics of card, plastic, wood, paper and polystyrene, knew what would happen to these materials if they were squashed, had some idea about their relative strengths and had begun to explore how to make strong, stable structures.

One step towards enabling girls to view themselves as scientists may be to find ways of validating their understanding of scientific concepts by helping them, and boys too, to recognize that scientific concepts are applied in a range of contexts. Helping girls recognize their scientific understanding within applications where they have confidence, gives these activities status and also provides a basis for enabling them to transfer their knowledge to other contexts which may lie within 'male' domains.

It is widely recognized that children may have an understanding of a concept within one context but this understanding will not be apparent if the child is asked to apply this concept in another context (Donaldson, 1978; Hughes, 1986). We suggest that the existence of gender-related domains can be a major factor in influencing how well children can apply their understandings and capabilities. Although it is well documented that girls do not use constructional and mechanical toys as often as boys, they certainly have the necessary understanding to perform competently in spatial and mechanical activities and have developed a range of skills associated with CDT. Many 'female' activities (e.g. making dolls' clothes, dolls' houses, paper baskets) involve the process of conceptualizing a finished product, considering how to construct it from a range of non-specific components, executing a plan and modifying it as necessary. The same process is involved in building something from a construction toy. It seems reasonable to assume that the use of construction toys is influenced by senses of entitlement and confidence, rather than simply about possessing a particular type of knowledge and aptitude.

One approach to helping girls gain access to scientific application within male domains which we have found to be relatively successful is to begin with activities with which girls are familiar and feel confident about and move gradually into situations where they can apply their abilities to less familiar contexts. For example, an exploration of movement with five infant girls began with drawing animals, making movable paper puppets of the animal and finally making a model out of the construction toy Mobilo, a resource which offers scope for the exploration of forces and movement but one which these children had refused to use up to this point. Helping these girls transfer their understanding and interests into realms they defined as 'male' (e.g. construction toys) enabled them to extend their use of materials and begin to question the 'ownership' of materials and resources.

There is no simple way of promoting equal opportunities in science, maths and CDT. The observations outlined in this chapter highlight the complexity of the issue. Children's attitudes towards science and technology reflect deep-seated social patterns and therefore cannot be modified by a single, simple strategy. Real progress is dependent upon changes occurring in society as a whole. However, teachers cannot wait for such changes to occur and, within the classroom, teachers can have some impact on children's developing attitudes, confidence and choices. In order to do so we need to adopt a range of approaches which aim to promote equal opportunities. For example, ensuring girls spend more time with construction toys should be seen as only one way of dealing with a multifaceted issue. Approaches must also address other aspects such as attitudes towards various activities, the way activities are defined and identified, the degree and nature of adult involvement and methods of assessment.

This chapter focused on children's behaviour and comments in order to highlight a number of issues about the way in which girls operate within the classroom. Subsequent chapters explore practical strategies for effecting change.

Note

We would like to thank the children and staff of the 29 schools in which we observed and worked.

CHAPTER 5

Fair Play: Children's Mathematical Experiences in the Infant Classroom

ROSE PARKIN

In both early-years science and early-years maths the emphasis is now on investigational and problem-solving activities – certainly of action upon real objects. It has long been suggested that girls do not perform well in mathematics because of a lack of meaningful constructional activity which hinders the development of spatial ability. Many existing explanations of the difference in performance between girls and boys emphasize this within early play. Byrne (1978) suggests that early experience with constructional toys helps to underpin spatial awareness which helps future mathematical understanding. This explanation could be rather damning for girls who are either denied this experience or who are less likely to choose constructional equipment during spontaneous play. Could there be an overvaluing of constructional play and an undervaluing of the play which girls are more likely to choose? Even if boys do perform better than girls in spatial ability tests (Chipman and Wilson, 1985), Shuard (1982) questions why spatial visualization is emphasized rather than verbal ability, in which girls do better than boys, when it is not yet known how far either relates to the learning of mathematics.

Other questions now arise. Do girls really lack spatial ability? Many teachers would answer yes, but is this based on traditional belief or on classroom observations and assessment? Does experience with construction toys deserve the importance it is given? Do girls develop spatial ability in other ways? My own experience has indicated that girls can perform well in investigational work and problem-solving if the groups are sensitively organized and the problems are not dominated by traditional male-role life situations.

Instead of looking at the difference in girl–boy constructional play alone, I was keen to investigate other three-dimensional activities which might help girls develop spatial concepts. I also wanted to look at girls' behaviour when choosing freely and analyse the mathematics in their play.

The research so far

Before reporting the results of my own observations of infant mathematics activity it is important to consider existing research concerning the difference in classroom behaviour and performance between girls and boys. It seems very likely that in many aspects of school as well as social life and later working life, girls learn that it is best to become 'invisible' (Spender, 1982). Girls find that keeping a low profile is desirable, beneficial and certainly expected. They seem to learn that drawing attention to themselves is not usually considered acceptable behaviour. In some intellectual pursuits this might aid development, e.g. reading, imaginative writing, pattern making, computation. However, in other areas (and many of these would be mathematical), this behaviour would hinder development, e.g. activities which involve large equipment, organizing it, moving it, working together with others and coming to decisions about its arrangement.

So why do girls tend to become passive in their learning environment while boys seem to become active and develop confidence? And why do some girls begin to develop negative attitudes to mathematics? It is necessary when considering socialization and its effect on girls' behaviour, to look at influences both inside and outside the classroom. Historically this is not a question of mathematical performance alone but a general attitude to the educational needs of girls. There is much evidence to show that this passive behaviour of girls in educational situations has been known and its dangers recognized for a long time.

A teacher of girls is, perhaps, too easily satisfied when her pupils are working steadily and conscientiously along the lines which she has laid down for them; a boy is almost certain to go off at a tangent . . . routine does him less harm, because he is less susceptible to its influence. Probably one of the weaknesses of girls is that they will submit to so much dullness without resentment. . . . (Gwatkin, 1912)

Weiner (1980) found that boys receive more attention, punishment and praise, from adults than girls. Adults respond to boys as if they find them more interesting than do girls. Therefore it seems safe to say that girls will see themselves and their ideas as unimportant and so learn to take a 'back seat'. It is interesting that Walden and Walkerdine (1982) found that teachers considered girls far easier to assess than boys whether they were considered of great or poor ability. Again, boys are thought more important, their behaviour more profound and deserving more careful assessment than that of girls.

Eynard and Walkerdine (1981) report that their studies of spontaneous play, showed that girls chose creative and imaginative play while boys chose construction toys. But is this spontaneous behaviour of girls really the long-term result of being pushed out by boys? During my observations I found that girls complain less than boys do if they are denied a chance to participate in constructional activities. Maccoby and Jacklin (1975) noted that boys are given more spatial and scientific toys in the home environment and certainly boys seem more likely to have access to calculators and computers at home than girls. Perhaps girls' play patterns are attributable to girls opting for the familiar.

Dweck *et al.* (1978) found that even positive and negative feedback is different for girls and boys. Negative feedback to boys mainly concerned their conduct while negative feedback to girls mainly concerned intellectual aspects. Fennema (1981b) found that boys tend to attribute their failures to lack of effort and their successes to ability whereas girls attribute their failures to lack of ability and their successes to their effort. The APU (1979) similarly found that at 11 years old girls were already attributing failure in mathematics to lack of ability while boys displayed much more confidence.

If, as seems likely, these attitudes to girls and the attitudes of girls have existed for generations, then it would follow that parental attitudes to the mathematics education of girls is not encouraging. If a parent's (particularly a mother's) own experience of mathematics was unpleasant and the subject was rejected, it is possible that her expectations of her daughter will be low. This expectation would be reinforced further by the commonly held belief that it does not matter, that mathematics is not necessary for girls because women's work does not demand it. When I expressed concern about one girl's poor performance in mathematics her mother replied, 'Oh – I was like that'. Would she have said the same had it been her daughter's reading that was causing concern? I suspect that men and boys would be less likely to admit to failure in mathematics.

Walden and Walkerdine (1982) mention the 'catch-22' situation that girls find themselves in.

> If they fail at mathematics they lack true intellect but are truly female. If they succeed they are only able to do so by following rules and if they conquer that hurdle they become somehow less than female.

There is also a lack of role models in the form of women mathematics teachers in secondary schools. I was once told by an LEA mathematics adviser that primary girls do not have the same problems of positive gender identity in mathematics as do secondary girls because most primary teachers, teaching across the curriculum, are women. I found this rather shortsighted as in my view women primary teachers are just as susceptible to environmental influences as any other woman and just because they teach maths it does not follow that they present a positive role model. Aiken's (1970) research indicates that women student teachers feel that their male peers are more mathematically competent than they are.

The maths education of girls and boys might not on the surface be significantly different but attitudes to the mathematics education of girls still has great influence. The National Curriculum might make us all more aware of the need for equality of opportunity and the assessment of individual development, but will testing take account of different girl/boy mathematical behaviour and interest and will these differences be fairly valued?

Classroom observations

My own observations of reception and middle-infant children were made over a period of five terms at a school where the vast majority of children were from a Muslim background and English was not their mother tongue. I began observing the 'free play' of my class of 27 mixed reception and middle-infant children; 17 boys and 10 girls. The teaching medium was English and the ability of the children to communicate in English varied enormously and although on the whole girls seemed to become competent English speakers sooner than boys, the boys developed other communication skills at a faster rate.

The girls tended not to display the same confidence as did the boys. There was no doubt that boys dominated activities in that classroom, through their numbers and confident assertiveness they were able to control much of the 'free' activity of the class. This is not to say that there were no confident, assertive girls and no boys needing support and encouragement in order to socialize; there were both. Yet proportionately more boys than girls displayed, both verbally and physically, a determination to control activities which verged on aggressive behaviour. Girls, on the other hand, when unable to participate in the activity of their first choice, accepted their second or third choice without showing resentment. Under such conditions it was difficult to observe 'free' choice and come to any meaningful conclusion. If boys were found to be involved mainly in construction activity and fantasy play and girls mainly involved in the home-corner and pattern making, it did not necessarily mean that girls had made those activities their first choice, merely that boys had not chosen them.

It was in this situation that I first considered limiting the choices for boys. This was not easy in a small classroom where resources and space were not abundant, and where, for most of the time I was teaching alone. Because of the particular needs of these children, working with groups of more than four was impractical. I decided to limit boys' choices for just one hour each morning. For part of that time they were completing set tasks e.g. a practical problem, reinforcement work on sheets, table-top games, following taped story in books, using the computer in pairs. For another part of the time they were receiving direct teaching from me either individually or in small groups. For the remainder of the hour they would be choosing from a limited selection of activities. I attempted to arrange these one-hour periods to coincide with a welfare assistant being with me but as this was not always possible it was vital that the boys' activities were not too demanding yet at the same time sufficiently stimulating to sustain their interest for 20 minutes.

It was very hard to establish this change in the routine. At first the boys were confused about the limitations or would not accept that some things were out of bounds. For myself, insisting that children be limited in their movements for such a length of time somewhat went against the grain. However, these discomforts were insignificant when measured against the obvious benefits to the girls.

I provided as much construction equipment as possible, both large and small, being very much aware of the current belief in its importance in the development of vital mathematical concepts. There was always sand or water (very often both), the home-corner, dressing-up clothes, the book-corner including large books for group reading, painting, some other creative activity or dough and a table for writing or drawing. At first girls were hesitant with activities they had had little experience of. They moved from one activity to another without stopping long enough to produce any meaningful results. After a few such sessions, the girls began to explore the equipment in a way they had previously been unable to, and at this juncture I felt that making observations of the movements of boys and girls could be useful.

Girls continued to approach three-dimensional construction activities very tentatively. It was very common for them to attempt to make containers of different shapes and proportions and to use smaller pieces to fill these containers. Perhaps not the most imaginative use of the equipment but interesting and mathematical all the same. Sometimes these containers would be musical instruments, shakers, and they would sing while playing. Other than the simple box structure, towers were a favourite, also tubes, bridges and tunnels. During girls' co-operative construction they used mathematical language to describe, direct and plan. This is important, especially the use of prepositions, which in my experience is the last part of speech to be learnt and for those learning a second language it comes even later. Building, using equipment as a framework rather than walled containers was rarely practised, nor was the making of wheeled vehicles, except sometimes houses on wheels.

I was interested to find that girls very often built abstract patterns rather than trying to make a likeness of a particular building, vehicle, etc. Only my questioning led the girls to ascribe a possible function for their creation. Symmetry of both colour and shape was an important feature to the girls. Some would search for much of their time to find the right colours to produce their models, and if their models looked incomplete it was usually because there were not enough pieces of the right shape or colour to complete the construction and second best would never do.

At the sand and water trays, again I found the girls preoccupied with filling containers. There was little building of sand mountains with tunnels running through or trucks leaving tracks through rough desert terrain. They enjoyed making shapes in wet sand using moulds, and making sand 'food' to put in dishes and cups to share out. They investigated all the possibilities of consistency, adding water a little at a time and stirring with spoons. With water, again, there would always be the sharing game, filling bottles and tea cups, but all the other water toys were also used. The girls were also involved in creative activities and the home corner was never empty. In the imaginative play area the girls would usually remove all the utensils from the cupboards and replace them differently. The home corner furniture would also frequently be rearranged. This repositioning of equipment was always a most important part of their play. The girls were skilled at manoeuvring this large equipment into the most appropriate position, in empty-ing and filling cupboards and in laying tables for varying numbers of people both practically and aesthetically. The mathematics in their play including three-dimensional construction, spatial visualization and planning was extensive.

Boys had free choice throughout the day, except for the hour set aside for girls, and during the time I was working with them directly. It was unusual at first for girls to compete with boys for a place at any construction activity although eventually some girls developed a preference for this play and became more insistent. When choosing alongside boys most girls opted for their traditional places, i.e. pattern making, home-corner, reading, sand and water, dressing-up, creative activities.

Boys differed from girls in the way they played and their use of the equipment. Boys used construction toys more confidently and purposefully. They made things, working independently or co-operatively they produced wheeled vehicles, boats, spaceships, spacestations, buildings with rooms, etc. They also worked for longer on their designs than the girls did and played with the result. They worked co-operatively, in pairs or small groups more so than the girls who seemed to prefer to work alongside each other but on individual designs.

Boys used the home-corner but not for the same lengths of time as girls and most boys used it differently to girls. Some of the boys joined with girls in their home-corner play, but boys playing together tended to transform the home-corner into a base for some broader fantasy, e.g. a fire-station, for some greater drama being acted out in another part of the room. They were constantly in and out and did not use and arrange the equipment in the same way as the girls. Sand and water play was as popular with boys as with girls, but boys used trucks and tractors more. They were also more interested in sand and water wheels and in plastic tubes at the water tray. The boys engaged in creative activities in much the same way as girls.

I continued to organize my day in this way for a period of nearly five terms. During this time the composition of the class altered. Some of the children remained with me for the whole time, but most of those I had started with had moved on by the end of the period and new children had joined the group. There continued to be more boys than girls. Occasionally I stopped 'girls' hour to help me judge its value, and although it seemed that girls had become more confident in choosing equipment when competing with boys, I observed that their concentration span was shorter, that boys constantly 'interfered' with girls' work (sometimes wanting to help, sometimes wanting to take over), and that girls would 'give in' easily and revert to their traditional classroom activities. I decided to keep 'girls' hour not only because it gave girls the freedom to explore and use construction equipment and other classroom activities in their own way without hindrance, but I realized how important it was for boys to see girls working confidently in these areas, to accept girls' right to make construction play their first choice and to appreciate the value of girls' work. As far as was possible designs were discussed by the class, comparisons made, and the children's attention drawn to their size, shape, colour, symmetry, proportion, and to their function or aesthetic value. The children were also encouraged to draw and write about each other's models. In this way one child's construction could stimulate another child's investigation, and children learned to see the satisfaction of a purely abstract model or the practicality of true-to-life representation. No type of design was given greater value.

It seemed that it was the different classroom behaviour of girls and boys which determined where they spent most of their time, and they were not necessarily, especially the girls, exercising real choice. Different classroom behaviour leads to different classroom activity and this might lead to the development of different skills. Different classroom behaviour seems also to have either a positive or negative effect upon participation in particular learning methods.

As mathematics co-ordinator in this same school I was particularly interested in developing investigational and problem-solving activities, and began teaching my own class in this way early on in this five-term period. Both investigation and problem-solving demand perseverance, concentration over time and, as much of this work is carried out in groups, a degree of assertiveness. Girls performed well during individual investigations, e.g. find as many ways as possible of arranging five squares joined on at least one side, but performed less well when involved in mixed gender group problem-solving. It was not only that boys' assertiveness intimidated the girls causing their lack of concentration, but girls also tended to drift into individual activity or paired work even within a group. They would often refrain from discussion, sharing a point of view, disagreeing or demonstrating. Boys on the other hand tended to offer an opinion and were prepared to take risks. This difference between boys and girls showed itself both in single and mixed gender groups but girls participated more actively in all-girl groups. However, even in this situation girls had a tendency to work alongside rather than with each other.

I observed a problem-solving session involving a group of four girls and boys. They were given two play people and asked to build a wall taller than one play person but not so tall that one play person could not see over it standing on the other's shoulders. Having built the wall they were to build a staircase up one side and down the other. There was a quick discussion about the best blocks to use and the girls began to sort by shape and size. They rejected cylinders but kept triangular prisms and other shapes with three or more flat surfaces. One girl insisted that the triangular prism could be for the top. The boys had also sorted a pile of blocks mainly large cuboids and were already laying the base. Boys and girls continued to work separately laying the blocks at different ends of the wall. The boys conversed loudly and constantly about the stability of the wall and also tried to direct the girls. They were preoccupied with the structure and needed to be reminded about the requirements of height. The boys seemed

irritated with this restriction whereas the girls seemed glad of the respite, taking the play people, measuring them against the wall and indulging in imaginative play, giving them voices thereby providing more purpose to the solving of the problem. The boys joined the two ends of the wall together (they were unequal in height) and the girls added decorative top shapes. The stairs caused more division, vociferous argument about a base, the height of each step, the height of the staircase, etc. The boys dominated this discussion and the girls drifted away. The boys completed the staircase alone. Although the girls' section of the wall was in no way inferior to that of the boys, they lost confidence in completing the task. The boys seemed to expect to solve the problem. The girls were either not sufficiently interested and therefore prepared to let the boys take over, or they also expected the boys to do better and withdrew. The latter seems most probable as the girls' interest at times seemed strong. This pattern of behaviour was repeated frequently in mixed gender group problem-solving.

During other, usually individual, mathematics activities girls performed as well as boys and sometimes better e.g. positional activities such as placing the brick in/on /in front of/behind/under/next to/ the box; activities involving one to one, one to many, many to one correspondence; tessellation using regular plane shapes; making sequential patterns using colour, shape or both.

Observations made by a colleague in classes of middle infants showed that when children were directed, differences in behaviour became more obvious, e.g. when girls were allowed only construction activities and boys allowed only home-corner, drawing, Plasticine, painting, story writing, shop, puzzles and small games, 50 per cent of the boys were happy while the other 50 per cent complained loudly and frequently. Eighty per cent of the girls were content while 20 per cent lost interest after approximately 20 minutes but did not complain out loud.

When choosing freely, girls who were participating in a construction activity were less likely to show their creations to the teacher whereas boys did so constantly. Boys interrupted the teacher (reading with child) constantly to show their models. Girls did this far less frequently.

Teachers I have spoken to are beginning to put restrictions on various activities because they are becoming aware that girls enjoy some construction activities but are unlikely to compete with boys.

Observations of top infants show a greater division between the choices of boys and girls. A top-infant class was observed over a five-day period. The whole class had been involved in plane shape identification, tessellation and pattern as a subsidiary activity. It was a revision theme and enjoyed by most children. There was a tendency for girls to return to it more often than boys. More girls than boys were involved in large-scale tessellation projects on paper, i.e. involving colouring around drawn shapes or reproducing patterns with sticky paper (not pre-cut shapes). Girls and boys showed equal interest in making the patterns on the table top with plastic shapes. Where patterns were made of several different shapes, there appeared to be no difficulty for either girls or boys to 'fill the gaps' with an appropriate shape.

As an extension, the activity was to use Polydrons to build something. They looked at the pieces available, talked about appropriate pieces to use as axles for the wheels, how pieces fit together, whether they stand up to make a tall structure, whether they thought they would be able to make a box, etc. Having been aware from previous experience that construction activities tend to be dominated by boys, the teacher chose a group of five girls to have the first turn with plenty of space on the carpet, no interference (the rest of class occupied elsewhere) and three crates of pieces. The girls fingered the shapes then sorted by colour for a short while. They fitted the smaller shapes into the larger frames to make two colours, then decided that only yellow was suitable for centres (obviously making flowers) and lost interest in that activity as there were not enough yellow shapes. There followed a half-hearted attempt to fit yellow uprights on to flat shapes but this was not very interesting. They then made tessellation patterns with squares and hexagons. This activity lasted for 75 minutes. The teacher did not intervene or become involved but sat nearby reading with other children.

On the second day another group of four girls were given the same task. They made tessellation patterns immediately but soon lost interest and went on to make dumbells with yellow tubes and two wheels and were quite interested for about 10 minutes trying to make the centres longer by joining several yellow tubes. The teacher intervened and asked 'could you make the weights heavier?' – this question was not taken seriously. They finished by sorting plane shapes according to shape. The activity lasted for one hour.

On the third day the same task was given to a mixed group of three girls and two boys. The boys immediately began making a box structure, a lorry. The girls did not participate but watched and then decided to make their own. Boys and girls completed models separately. Boys had used triangular shapes to obtain a cab shape, they appeared to have no difficulty in seeing which shapes were needed. The girls' lorry was a rectangular prism with four wheels on two axles. With the teacher participating they found that the roof was not stable in either model. The girls showed little interest in redesigning this. The boys, however, tried several strategies, discussed the problem and were prepared to let one of their group rectify it. They returned to evaluate, someone else had a go and all were involved in the problem and solution. Interest was maintained for 75 minutes.

On the fourth day this same activity was offered for free choice. No girls worked with the Polydrons that day but a core group of boys played co-operatively while others came and went seeming to be more interested in monitoring the activity than joining in but they added the odd bit or made suggestions that were well received. There were no disagreements. On another table the teacher had made shapes and sticky paper available as a free-choice activity. Successful outcome with good concentration was shown by all participants – all girls.

These observations of infant children have led me to the following conclusions. The free choice of young infants does not differ significantly between girls and boys but there is a difference in the way boys and girls behave and in the ways they use resources. Girls' behaviour prevents them participating equally with boys in many problem solving and other practical classroom activities but their behaviour is advantageous to other expressions of mathematical concepts. In certain mathematical activities girls perform as well as and sometimes better than boys. Positive discrimination in favour of girls can increase their interest and confidence in constructional play with no detrimental effect upon boys.

Girls' poorer performance in certain areas of mathematics should be viewed in the same way as girls' and women's underachievement in a variety of situations, i.e. either the same opportunities are not open to girls and boys or that it is somehow considered inappropriate behaviour for girls. These causes are either con-sciously or unconsciously reinforced by teachers, parents and peers. These pressures on girls increase as they become older and might also differ according to social class and race. Such attitudes create a division in children's experience of mathematics where girls tend to be passive and boys tend to be active participants in their learning.

School Accounts and Action

The Role of the Nursery School in Developing a Non-Sexist Approach to Science and Technology

EVE LYON

Even at the age of 3 years, children come to school with gender roles firmly in their minds. Furthermore, whether or not we admit to it, these gender roles can be reinforced by school staff and the school curriculum. Girls are stereotyped as passive, emotional and unpractical whilst boys are expected to be noisy, active, messy and interested in constructional and technical activities. The signals we give children can accentuate these stereotypes or can positively attempt to challenge them. The early experiences of a child's life can be vital in determining her or his later attitudes and expectations. The issue of sexism in nursery schools therefore ought to be given a high profile. Challenging long-established prejudices is a difficult task but it is essential to rethink our practice in nursery schools if any real progress is to be made in encouraging pupils to widen their horizons.

I chose to focus on science as part of our school development plan for two reasons. Firstly because development of the skills and thought processes encompassed in the learning of science are vital for all young children. Secondly, science is an area where bias and stereotyping are particularly obvious. In focusing on science we aimed to give very young girls a positive start in an area which is later biased towards, and dominated by, boys and men. I also believed that a focus on science in conjunction with a concern for gender equality would strengthen our determination to provide an environment and a broad curriculum in which every child, regardless of sex, class or ethnic background, would be treated as an individual and become more independent in their learning.

Our work on developing science coincided with the introduction of the National Curriculum and the publication of the Final Orders for science. Our school welcomed the National Curriculum for science for a number of reasons. To begin with, our children experience many of the programmes of study included in the National Curriculum as the learning processes involved in science do not begin at 5 years. In addition, the introduction of the National Curriculum provided an ideal opportunity to extend our already positive links with the neighbouring infant school and enabled us to put continued emphasis on a developmental approach to learning central to our school's philosophy and to the ideal of early-years education as a continuum from 3 to 7.

In order to develop science in our schools we had to tackle a number of issues: motivating and involving all staff, organizing the classroom and equipment to encourage all children to participate equally, informing parents and perhaps challenging parents' attitudes and embracing the National Curriculum in such a way as to promote and value the early stages of young children's learning in school.

In my second year as Head and as we approach the end of our first year's work on the project, what initially seemed to be a dauntingly mammoth task is slowly becoming part and parcel of our everyday concerns. Staff whose first response was apprehension and unwillingness to become involved now use the word 'science' freely and make observations on the girls' achievements in a positive way. New staff and visiting students are not able to ignore this strength of commitment. Parents, and not just the small dedicated body which is a part of every school, regularly arrive with old clocks, batteries, old radios, etc. Parents' initial surprise that these may be useful in their child's education has gradually changed to confidence and pride, and some actively seek out things for the children to use. Most heartening, and of course crucial, is the fact that the children themselves demand that specialized equipment is provided and that they can use this equipment themselves.

Motivating staff

Staff in the early years have always been confident about most aspects of natural science but nervous about approaching more technical topics and problem-solving activities. Most of us who are involved in early-years education are women and we bring to science our early experiences and with them our inhibitions and reservations. Very young children, on the other hand, are often very attracted by the unknown and the unfamiliar. We wanted to build on this natural curiosity and on their everyday home experiences in the hope that by the time they left the nursery they would approach science with confidence and determination. To achieve this all staff, teachers and nursery nurses, needed to believe that science for young children was a valuable part of the school curriculum and also to believe in their own ability to actively encourage and support children's learning of science. It would have been relatively simple to produce a science policy booklet for staff and parents outlining our objectives, organization, topics and equipment. Science could have become a neatly packaged addition to the curriculum but as such would have been of limited value. I wanted more than that and was keen to build on the child-centred approach already strongly evident among staff in planning and organization.

Several years ago, as Deputy Head, I was asked to take responsibility for science in my school. I had an arts degree, was a woman in my mid-thirties, had taken little interest in science and was somewhat in awe of it. I went on a series of science courses at the local Teachers' Centre – most of the teachers were women from the primary sector – but the tutor was a man. I was slightly intimidated and ready to sit back and let other colleagues teaching older children do the work. This did not happen. The tutor began by talking of his experience and observations in nursery and infant schools. He obviously valued the work that was going on in nursery schools and classes, he mentioned the lack of any structured progression from 3 to 7 in science. I thought of all the things which went on in my school in cooking and water play, in the experimenting children did with growing things, and how these stimulated children's thought, language and problem-solving skills and felt encouraged, even confident. It was a different matter when we moved on to a workshop session – electricity. We had to use real equipment which I found unnerving and scary. For a short while there was silence but not for long. Soon, in pairs we handled, touched, experimented, observed, talked and even succeeded in our investigations. It was for me quite simply, exciting – a revelation. I felt childlike and for all my teaching experience and expertise I perceived, in some quite basic way, what it may be like to be a 3-year-old growing, exploring and finding out about her or his world. It is this revelation which has stayed with me and encouraged me to make science an integral part of nursery life. I don't know if a male teacher would have felt the same, or one from a different class or background, but I was aware that a lot of my lack of confidence stemmed from the fact that I was a woman with quite stereotyped attitudes about my abilities. Long before the courses were finished my attitudes had changed drastically and with this my confidence. I took many ideas back with me to the classroom – I wanted the children, girls and boys, to experience some of the thrill of discovery I had encountered, and indeed, the addition of 'science' as an integral part of the nursery curriculum seemed a natural extension of many problem-solving activities children were experiencing in all areas of play. From the earliest stages of life all children, regardless of sex, race, class or disability, are solving problems. The young baby attracts a parent's attention, the toddler manoeuvres a wheeled toy across a crowded room, the 3-year-old builds a bridge or construction from bricks. Science takes these learning processes and extends them, introducing specific skills and language, close observation, experimenting, hypothesizing, testing and using particular equipment. The later imbalance between girls and boys in their experience of science was an issue I felt I had to address and as Head of a nursery school it seemed natural for me to introduce science in the way my tutor had.

Getting started

We began our in-service with a 'brain storm' with everyone involved, every suggestion was written down, prejudices and concerns as well as positive points. This was followed by a series of workshop sessions using everyday materials already available in the nursery – paper of all sorts and sizes for folding and magnets with different materials and mediums. We investigated simple play activities like bouncing balls, and we also used more technical equipment. It was a whole team approach of mutual support and encouragement and it was fun. The school has a male Deputy Head who is committed to equal opportunities for all children and a young nursery nurse who is particularly enthusiastic about extending the children's experience of physical sciences, design and technology. Such was the value and impact of our staff workshops that all staff, however dubious at the outset, became interested in science and, as women, were aware that much of their lack of confidence in this area was due to sexual stereotyping. Consequently, they also felt quite strongly that equal opportunities in terms of gender, should play a significant part in the development of science throughout the school.

Putting our plan into action

Our next step was to put our plan into action. The school is dedicated to encouraging the children to become independent learners. All the equipment and materials are stored to allow access to all children and they are encouraged to take responsibility for it. If additional materials are required in any area of their play they are free to ask for them and go with a member of staff to the central store. Topic work comes from the children's interests, from the local or school environment or from a wide variety of outings and visits. Creative activities are based around a workshop organization and with an extensive variety of paints, glues, collage materials with the woodwork bench immediately accessible. All children are actively encouraged to experiment. What may begin as a painting or collage can finish as a model incorporating wood, nails, glue and junk boxes. The feeling that anything is possible for the children is central to the school's philosophy and, as far as is practical, children and staff work as partners in planning for activities and play situations. If the home-corner is to be converted to a hospital or shop the children are involved in the planning and design, trying things out, talking all the time about what is involved and what may or may not work. Discussions and a system of trial and error are implicit in this approach.

There had been an abundance of natural science going on throughout the school and we looked at ways of developing skills of careful observation and practical enquiry, encouraging relevant questioning and discussion in these areas. When planting seeds in different situations and under different conditions we wanted to develop an awareness of variables influencing results and enabling children to become confident about predicting and hypothesizing. Our children regularly make personal books using our book-binding machine and with photographs, drawings and writing they are able to record their findings clearly and are encouraged to communicate these effectively to others. This emphasis on particular 'scientific skills' in an area where staff and children were relaxed and confident provided the springboard for the introduction of more work in the areas of technology and physical science. The school was awarded a grant of £300 from ILEA's Gender Equality Fund for work with girls and science. Some of this money went towards the production of a photo booklet for parents which we felt was an essential component of our plan. Involving parents, seeing them as partners in their children's education and keeping them informed of new developments in school is always given a high priority. We felt it to be particularly important in this area as gender stereotyping is evident even at the age of 3 and raising parents' expectations for their children, particularly the girls, was absolutely vital. Our booklet outlined the work the children were involved in at school using clear photographs, displays and captions, under the title of 'girls can do anything – so can boys'. Our focus on science and equal opportunities coincided with the publicity about the National Curriculum in infant schools and this provided us with an ideal opportunity to make parents more aware of the value and importance of the early stages of their child's experience and learning. Consequently, parents' contribution to our science resources have been invaluable. Science constituted the major part of this booklet with captions such as:

> Natural science encourages children's appreciation of the environment around them and develops close observational skills.
>
> Problem-solving and experimenting are the first steps in developing an interest in science.
>
> Sand and water play enable children to develop an understanding of early mathematical and scientific ideas as well as being both enjoyable and rewarding.
>
> Construction can be challenging for both girls and boys – it is harder to make a car or videocamera than a gun or sword and

many problems can be encountered and overcome. Construction play also requires and encourages discussion and co-operation.

We also wanted to place science in a whole school philosophy of equal opportunities for all children so we included photos and captions on other areas of the curriculum:

> Children can explore making their own music using instruments from different cultures.
>
> The children enjoy the experience of a variety of cultural festivals.
>
> The familiar activity of cooking enables children to enjoy foods from different countries and in the process they experience a range of mathematical and scientific ideas.
>
> Physical development is an important part of the nursery curriculum. Non-slip shoes and practical clothing can enable all children to take part.
>
> Parents are always welcome and are encouraged to be partners in their child's first experiences of school.

The remainder of the money was used to buy basic scientific materials – magnets, bulbs, batteries, magnifiers and general equipment. As with our other resources these were stored in a way to allow access for all children. Some of the equipment, such as electricity sets, required quite specific introduction by an adult and all are required to be used and cared for properly. All our children are encouraged to value the resources of the school – they know for example that books need to be respected and treated carefully and that hammers, saws and woodwork equipment can be dangerous and must be used with discretion. Our approach to science equipment was to follow the same lines. We planned to ensure equal access for girls by positive encouragement from our mainly female staff, with the involvement of mothers in experiments and activities, by careful observation and record keeping and if necessary by positive intervention.

As part of our work with the Equal Opportunities Working Party our school became actively involved in an ambitious project for Women's Week at the Teachers' Centre. Workshops for craft, design and technology and alternative provisions for the home-corner were open for use by visiting schools. We sent children to these workshops and our 3- and 4-year-old girls approached tasks and equipment with a confidence which impressed many primary colleagues. We returned to school with the loan of a 'working garage', with real equipment and tools – this role play situation so often associated exclusively with boys is rarely seen without at least an equal quota of girls, and often mothers, fully dressed in overalls and masks.

Much of the craft, design and technology work that the children now enjoy has come from the extension of their workshop activities. Every child's efforts and models were valued and staff encouraged children to try them out. Adults asked the children to think critically about the models – did the car with wheels glued to each end actually move or the boat made from paper really float in the water? Discussions between staff and children took place and ideas were developed and suggestions taken up. The construction of 'sailing boats' and one girl's detailed plan for a kite are just a few of the examples of the type of scientific investigations the children were involved in (Figures 6.1 to 6.3).

The children's achievements are always valued and displayed for parents and visitors. Girls and boys often show their work at infant assemblies and the nursery and infant staff are sharing meetings on the National Curriculum.

The staff at our school is committed to ensuring equality of opportunity for all children and, if asked, would say that girls are not discriminated against in any area of the curriculum. So, was our work on science and gender necessary or valuable? I would argue that our work has heightened our own and many parents' awareness of the importance of equal opportunities for girls in science in the early years of their schooling. In addition, our school is now providing additional learning opportunities in physical science and CDT.

It is a small beginning but nursery schools have a vital role to play in this field if bias in later life is to be avoided.

Figure 6.1 Instructions for making a kite from Kirby.

1 First you get some paper. You fold the corners down at the top, then you stick the corners down with some masking tape.
2 Then you do the bottom in a line and fold that down and cut the corners off.
3 After you have to put a tail on and a piece of string to the other side, and one underneath.
4 Then you draw a picture and write your name and stick some decorations on it.
5 Then just run very, very, very fast and it will fly right up in the air.

Figure 6.2

Figure 6.3

CHAPTER 7

Design It, Build It, Use It: Girls and Construction Kits

KIM BEAT
(on behalf of the
'Design It, Build It, Use It Collective')

We are a group of primary teachers (in Brent) who recognized our own and our pupils' needs to become technologically literate.[1] In particular we were concerned about the gender-differentiated use of construction kits (e.g. Lego, Mobilo, Big Builder, Rio Click, Ludoval and Sticklebricks) in the nursery and infant school.

Our group was formed as a result of a workshop on the use of construction kits. Our minds became focused on the kinds of problems we had when using kits in the classroom, and this raised a large number of issues many of which we felt unable to answer without further reading and some action research in our own classrooms.

This chapter will outline how we assessed the use of construction kits in our classrooms, why we identified a need for change in their use, the strategies we developed to encourage all children to take part in construction activities, how this can lead into many other design and making activities which are going to be an integral part of the National Curriculum, and how we then went on to develop a whole-school approach to the use of construction kits.

We all felt that the first step we needed to take was to assess what was going on in our own classrooms by observing the children and monitoring the use of equipment.

Assessing the use of construction kits

The first type of monitoring we undertook was during 'free-play' time. Using a plan of the classroom with activities available marked on it, we noted down, at five-minute intervals over an hour, which activities children were engaged in. Both gender and race were noted to enable us to see if any patterns emerged. The monitoring carried out in all the classes showed remarkable similarities. There was a general tendency for boys to gravitate towards the construction activities and, once there, to stay for almost the entire session excluding any girls who made attempts to join the group. Also throughout the hour the construction activities, which were usually on the carpeted area, would steadily spread to the surrounding area as more boys joined in. Meanwhile the girls were generally involved in drawing, colouring and cutting-type activities and in one particular case they were 'huddled' in one small area of the classroom while the boys dominated the rest of the space, sprawling along the floor and over tables. However, if the teacher engaged herself in the construction area then this encouraged girls to go and work there too. While the pattern for choice of activity according to gender was very marked no conclusion could be drawn from our monitoring to show that children's race affected choices during free play. However, in some classes children chose to work with others of the same race and this was particularly so where their racial group was in a minority in the class. Also bilingual children who were happiest working in their home language often chose to work with someone who spoke their home language. This is very important in enhancing self-identity and providing group support but it must be monitored carefully as it is important for all children to experience different models of language.

The second type of monitoring was much more detailed and involved close observation of groups of children engaged in construction play. We asked ourselves the following questions. What kind of models are the children making? Do some children always make the same type of model? What do the children do when they have finished the model, do they play imaginatively with it either individually or in a group? What type of language is used (imaginative, instructional, etc.)? Can the child explain how it works and how they made it? When a problem arises does the child give up, solve it by themselves or solve it with some help from an adult or child? Which children genuinely co-operate, sharing ideas and equipment? Which children dominate and which always choose to work on their own? Are there specific difficulties with some kits and how might these be overcome? This form of monitoring enabled us to build up a very detailed picture of each child's use of construction kits.

The general picture that emerged was that boys tended to make models that moved and were part of an imaginative scene they had created. They made such things as cars, fire-engines and spaceships which they talked about in great detail as they were making them and they almost always played with them when they were complete and would go to the teacher for praise and approval. Girls tended to make non-moving models such as houses or gardens and while making them they were usually talking about something quite different. When models were finished they were left, sometimes not even in a place where they would be safe. We also noted that when we worked alongside groups of girls their interest was maintained and they would respond to questioning about their model and make improvements. This impressed on us the importance of a positive role model being necessary if girls (and boys in other situations) are to break away from stereotyped behaviour. Many of our findings have been substantiated by similar action research.

Our third area of research was to find out how far pre-school experiences and parental attitudes affected interest in the use of construction kits. A colleague sent out a questionnaire to parents with a list of activities which go on in the nursery and asked them if they minded their children taking part in them. The list included such things as dressing up, playing with dolls, using Lego or tools, woodwork, speaking languages other than the home language or English, playing with water and sand, etc. This research was done in a racially-mixed predominantly working- class community. The findings indicated that most parents did not want their boys to play with dolls or do activities which they perceived to be 'girls' activities, while many parents did not want their girls to engage in activities such as woodwork or Lego – play which they perceived to be 'for boys'. However, there were fewer parents who minded their girls doing so-called boys' activities than boys doing so-called girls' activities. In this sample the race of the parent made little difference to the general findings. Another colleague sent out a second questionnaire to find out which children had experience with kits before they came to school and whether the kit had been bought for them or another sibling. This information was then related to the monitoring she had carried out when the children first came into the reception class. She found that kits had mostly been bought for boys in the family, and so, if a girl came from a family with boys (usually older ones) then she would have had experience with kits. Interestingly though, even when girls had played with kits, usually Lego, at home they did not necessarily feel confident to play with them in school.

Our research led us to the obvious conclusion that on the whole girls do not choose to play with kits during free play but when they do their play differs markedly from boys. This is the result of the conditioning we are all subject to from the media, our parents, our peer group and, of course, our teachers.

After playing with the kits ourselves we pooled together our ideas about the knowledge skills and values that children can learn when they use construction kits (Figure 7.1). In the light of the National Curriculum we related these to the appropriate Attainment Targets. This is by no means an exhaustive list but the exercise made us realize how important it was that we begin to employ strategies that encourage all children to take a full part in construction activities. However, we also realized that before we could encourage the girls in our class to use kits we had to feel more confident about using them ourselves. Some of us in the initial workshop had felt very nervous when it came to the 'hands on' part of the session – when we were asked to make something – so we all began to take home some of the kits from our classrooms to familiarize ourselves with them. We also had some sessions together where we supported each other when 'playing' with the kits.

The need for change

What we found was happening in our classrooms was greatly at odds with our sense of fairness but also with our philosophy of nursery and infant education. Our classroom practice is based on a belief in child-centred education within a structured and well-planned curriculum. It should be relevant, interesting and encourage children to explore and enquire while meeting their individual needs. All our work is guided by the principle of equality; the equal value of all people regardless of race, gender, class, disability

Figure 7.1 What do children learn when they use construction kits?

Knowledge

Size	M At8 Level 1
Weight	M At8 Level 1/2/3
Shape	M At10 Level 1/2/3
Symmetry	M At11 Level 3
Texture	Sc At6 Level 1/2
Cause and effect	Sc At6 Level 1/2
Similarities and differences	Sc At6 Level 2/3
Properties of materials	Sc At6 Level 2/3/4
Causes of motion	Sc At10 Level 1/2
Gears	Sc At13 Level 3
Levers	Sc At13 Level 3
Pulleys	Sc At13 Level 3
Colour	Sc At15 Level 1

Skills

Matching	M At1 Level 1/2 and At9 Level 2
Estimating	M At1 Level 3 and At9 Level 3
Counting	M At2 Level 1/2
Grading amd sorting	M At8 Level 1/2
Recognizing/making patterns	M At5 Level 1
Investigation	Sc At1 Level 2/3
Hypothesizing	Sc At1 Level 3
Experimenting	Sc At1 Level 2/3
Problem-solving	Sc At1 Level 2
Appraising	CDT At1 and 4 Level 1/2/3
Manipulation (making)	CDT At3 Level 1/2
Design	CDT At2 Level 1/2/3
Prediction	M At1 Level 1/2 and At9 Level 1/2/3
	Sc At1 Level 2
Measuring	M At8 Level 1/2/3
	Sc At1 Level 1
Comparison	M At8 Level 1/2
	Sc At6 Level 4
Planning	Sc At1 Level 2/3
	CDT At3 Level 1/2/3
Observation	M, Sc, CDT and Lang Ats
Communication	M, Sc, CDT and Lang Ats
Reporting back	M, Sc, CDT and Lang Ats
Questioning	M, Sc, CDT and Lang Ats
Research	M, Sc, CDT and Lang Ats

Figure 7.1 (continued)

Values	
Co-operation and sharing	CDT At1 Level 1/2/3
Coping with limitations	CDT At3 Level 1/2/3
Creativity/imagination	CDT At2 Level 1/2/3
Expression of feeling	CDT At4 Level 1/2/3
Perseverance	
Confidence	
Coping with success/failure	
Satisfaction	
Respect for others	
Self-esteem	

M = Maths; Sc = Science; At = Attainment Target.

and language. Therefore the issue of equality is central to the development of our curriculum and the way we organize our classrooms.

However, what we found did not show that our philosophy was being put into practice. Our so called 'free-play' situation was perpetuating inequality. Children were not free to choose for two reasons. Firstly, by the time they come into the nursery, children have already developed a strong gender identity which is more or less stereotyped. Even very young children have very strong ideas of what is 'right' for girls and boys to do (Belotti, 1975; Whyte, 1983) This can be seen in the activities they engage in, the skills and interests they have begun to develop and their interaction amongst peers and adults. Secondly, it is the teacher who channels the child's play and it is her set of values that determines what she considers to be appropriate behaviour for Black girls, White girls, Black boys and White boys. She decides on the activities in the classroom, she decides how and when to interact with children and when to approve or disapprove of different aspects of play.

Very often teachers see construction kits only as playthings and 'time-fillers', in other words construction activity happens when 'work' has been completed. This gives the impression to children that it is not important and therefore some children will not want to take part. Also we recognized, and admitted to ourselves, that there had been occasions when a child had brought their model for approval and we had said 'That's lovely, now go and break it up because it's hometime'. We certainly wouldn't have said the same to a child who had just shown us their maths, 'That's lovely, now screw it up and put it in the bin'! Our next line of enquiry had to be 'Well what are we going to do about it?' There ensued many months of trying out ideas and seeing which strategies worked best and in what situations.

Strategies

Our ideas have progressed considerably since those early days but the various stages of development are still valid starting points.

Classroom organization

It is important that construction kits are easily accessible to the children and that they know where they are stored and how they are expected to tidy them up at the end of a session. It is often more practical to transfer them from the manufacturers' packaging into plastic stacking boxes which are hardwearing and space saving. This can also overcome the problem of stereotyped images which appear on some of the packaging. Children will need to be encouraged to mix the kits, for instance Sticklebricks and straws combined can make some interesting models. Other media can also be used like Plasticine and play dough. Construction activities have to become part of 'work', not just something children can do when work is finished. Just as you would plan a language activity or a maths activity you should also plan a construction activity which children have to go to as part of their programme for the day/week. The teacher should always encourage children to finish their model as they would any other piece of work and provide opportunities for the child to talk about their model, particularly those children who previously did not take part in construction activities. Older children can be encouraged to design a model before making it and make modifications where necessary. Once the model is complete all children should be urged to play with it and perhaps make up a story to go with it. This gives children a sense of purpose and can guide them away from stereo-typical play. They should then be displayed with the name of the child/children who made it. Displays should always be attractive and varied with an interesting backcloth. Models made from kits such as straws lend themselves to being hung up and paintings, photographs and drawings of the models can be displayed around the table. With younger children the teacher could write down the child's description of the model to display with it. Older children can write their own. The teacher must ensure that she spends time with the construction group, as she does the others, offering praise and encouragement and asking questions to extend children in their work. Careful consideration must be given to the way children are grouped for construction activities. Our monitoring highlighted that boys often dominate construction activities so there may be times when it is necessary to have a girls-only group so that they can gain experience and confidence before working in a mixed group.

Getting started

Much of the research into girls and science, for example, Harding (1983) shows that girls are more attracted to problem-solving if it is put in a social context. The following ideas, some of which provide a social context, can be used as starting points.

1 Build a house for the play people. They will want a bedroom and a workshop. Try drawing a plan before you start.
2 Make something for the classroom pet, e.g. a playground, a transporter, a food dispenser.
3 Rupal made a scooter for one of the wooden dolls. Try to make something else that moves for them.
4 Make a model of a mosque, a temple, a church or a school. Think of a story to go with it.
5 There is a fire in one of the nearby houses. Make a fire-engine with a long ladder and the house that is burning.
6 Make a toy for the nursery children and show them how to use it.
7 Make a bridge to go over the railway to the park. How can you make it safe for everyone? Make sure that a person in a wheelchair can get over the bridge.
8 Make a model of the coach that we went in to the safari park. Now make some models of the animals we saw. You can use Plasticine and junk as well as kits.

To help stimulate ideas it is important that children have an opportunity to talk about what they are going to do before they start the practical work. The teacher can provide pictures and photographs which are applicable to the task.

Using photographs

While the children are engaged in construction activities it is useful to take photographs. These might be of completed models which can then be used on a taskcard to stimulate ideas or they can be used by the child to make a story or write a factual account of how they made the model. You can also take a sequence of photographs, for instance one girl made a fire-engine and the house that was burning and she told me the story of the fire. I took three photographs which she used to help her sequence the story and I wrote it for her. Then she told the story on tape, in Gujarati and English, so that others could listen to it and read the story. Your photographs can also show positive images of girls engaged in construction activities and of girls and boys working together co-operatively. These can be displayed with appropriate captions around the classroom. Photographs can also enhance a construction display or be used to make a class book about construction.

Encouraging co-operative working

Using construction kits provides the ideal opportunity for encouraging co-operative work. Our monitoring showed that more boys than girls found co- operating particularly difficult and this must be borne in mind when setting up groups. When children have not been used to working co-operatively it is best to start by working in pairs, paying great attention to the personalities that you put together. Obviously putting a dominant child and a quiet child together is not a recipe for co-operation. It is important to discuss with the whole class the ways in which decisions can be arrived at. Children need to be taught the language of co-operation. In the early stages of learning co-operative techniques the teacher must be prepared to intervene and talk children through the process when co-operation is breaking down. One strategy to develop co-operation is to ask one child to make a model, then to describe how to make it to another child and to help them as they build it. If at any time someone's model is broken by another child it is very important to discuss with them why it happened and to encourage them to help re-build it.

Moving on to design and making

As time went by and we became more confident about our own technological skills we began to feel that our work needed to be extended into other areas of technology. We also felt that it was important that this work became part of our structured curriculum not just something that was added on.

A very obvious way of extending model making with construction kits, is to ask children to recreate the model using a variety of junk materials. This also serves to preserve the image of the model which will eventually have to be broken up for others to use. Some children may want to make a model in junk first and then go on to use a kit.

As well as junk, pieces of wood, nails and glue can be supplied with a variety of tools. Using such materials will provide children with the opportunity to develop a wide range of manipulative skills. It is worth-while teaching children some techniques of construction after they have had some time to explore the various tools and materials. In this respect it is useful to have a 'tinkering' table for a while. Children are curious about the way things work. By collecting objects such as clocks, mincers, irons and cameras and supplying a variety of tools such as screwdrivers and pliers, children can actively engage in dismantling the various appliances. They should then be encouraged to put them together again! It is important to talk about safety and to impress on them that they should not take things apart at home. In one nursery, 'tinkering' was introduced to the girls first and then they introduced it to the boys. The girls were then seen as the 'experts' which really boosted their confidence.

By now we had come to the stage where we consciously planned to include design and making activities as part of our classroom projects. We started with the theme of Ourselves and below are some of the ideas that we included.

• *Birthdays* – The starting point was reading the poem 'Dilroy's Birthday' by John Agard. The children were asked to design and make a more appropriate card for Dilroy; a present using a variety of materials; a container to put it in and they were asked to design and print some wrapping paper. This introduced the notion of recycling of paper and some children made their own.

- *Families* – Children painted pictures of their families and wrote about them using the word-processor. This led to talking about homes, and children were asked to make a model of their own home which had an open front or top so that you could see the rooms which they made furniture for. Some parents and grandparents came in to talk about and show photographs of the homes they lived in when they were young. Many of the parents had lived in different parts of the world so this brought up the notion of homes and the materials they are made of being appropriate to climate, conditions and local resources.

- *Our school* – Children were encouraged to look at the appropriateness of technology within their classroom, the school or the local environment and to design a more appropriate replacement. For example in one school there was a toilet for the disabled but access could only be gained by stairs! The children measured the stairs and designed a ramp to fit over them.

Of course there are many more ideas which could be included in such a topic.

The National Curriculum

Design and making is an integral part of the National Curriculum and the use of construction kits, which have been outlined above, is wholly in keeping with the proposals for Design and Technology: Attainment Target 2. In the draft guidelines, entitled Design and Making, it is stated that 5–7-year-olds should

be given the opportunity to experience a wide range of designing and making activities, using a variety of materials. They should be involved in investigating, planning, designing, making, modifying and evaluating with appropriate guidance. They should develop or acquire further knowledge and skills when needed.

This was further emphasized in the programmes of study for Levels 1, 2 and 3 (DES, 1989d), where it was stated that:

pupils should develop design and technological capability through activities: based on contexts which are within their experience,

> both imaginary and realistic, such as children's pictures, poems and stories, the home, the school and local shops;

and further that,

> pupils should be taught to handle, use and explore a variety of materials and components such as fabrics, paper, card, 'dough-like' materials, disposable products and construction kits.

A project like the one described would provide ample opportunity for children to experience all the areas outlined in the above attainment target and the programmes of study. They were also only stating what many nursery and infant teachers have been doing for a long time!

Whole-school approach

In order for there to be continuity and progression in the use of construction kits, and for them to become an integral part of the design and making curriculum, there must be a whole-school approach to the use of kits. In order to achieve this we first had to convince our Headteachers that spending money on kits was a sound investment. Secondly we had to persuade them to allow us to run a series of workshops for the rest of our colleagues. We supported each other in the preparation of the workshops and spent a great deal of time trying to find approaches that would fully involve the staff and enable them to go through some of the processes we had gone through, but in a shorter period of time. Our aim was not to appear as 'the experts' but rather to be facilitators who posed questions and provided situations in which colleagues could arrive at their own answers.

There was certainly no lack of enthusiasm to introduce this kind of work into the curriculum. The biggest problem (as usual) was the lack of resources and in this case rather expensive resources. Here are some ideas that helped us to make the best use of the kits we had.

1 Pool kits into a central area to form a 'construction library'.
2 Pool kits between year groups or classes and arrange a rota for use.
3 Timetable hall-time for the whole class to use the kits.
4 Have a construction/science room or area for parents to work

with a small group of children.
5 Use construction kits in the playground, with mats to work on.
6 Put basic stock in each classroom and more expensive equipment in a library or rotate between classrooms.
7 Store kits in crates for easy movement. Ensure that children know the materials well and where they are stored.
8 Ensure a cross-curricular approach with theme work. Make use of the computer, tape-recorder and have feedback sessions to talk about models.
9 Involve parents and older children (girls and boys).
10 Combine kits to extend possibilities.
11 Use scrapbank, local factories or printing works to supplement materials.
12 Ask parents and community for any old kits that their children don't use any more.

Finally, I would like to describe briefly how two very different schools have been successful in organizing a whole-school approach to kits. The first school is a small infant school with about 120 children. The staff had for some time recognized the importance of construction activities within the curriculum but each class had only two or three kits. They found that the children had soon exhausted the possibilities with the kits and that enthusiasm declined. All the classes opened out onto a central hall area and it was decided that the construction kits should be pooled into this area to form a construction library from which teachers could borrow. It was also decided that one of the priorities for spending that year should be construction kits and that each class should have some basic equipment for its sole use and that larger and more expensive equipment should be added to the central library. Hall-time was allocated to every class for construction activities and, when possible, extra support was provided for these sessions either from language support teachers, part-time staff, willing parents or students. Many of the models were attractively displayed in this central area and children had the opportunity to talk about their models at assembly times. Also one of the top-infant classes was involved in a project to design a garden area in the playground. Once their ideas had been drawn together they worked alongside a bricklayer who taught them how to lay the bricks. During this project many photographs were taken and these were used to make a book as well as being displayed in the hall to provide positive images, particularly in respect of girls and construction.

The second situation was an infant department (250 children) in a large primary school. Two of the teachers in the school ran a number of construction workshops for the staff which resulted in an immense increase in the use of kits. There was also a commitment from the Head to spend capitation on kits and this money was used to 'top up' and diversify the existing kits. The building did not lend itself to a central resource area so it was decided to rotate sets of kits between classes in a year group as well as each class having some basic equipment. The kits were rotated on a fortnightly basis but there was always flexibility to borrow lots of kits for a day if anyone wanted to do a whole-class activity. As well as the emphasis on kits a number of teachers developed either a 'tinkering' table or a workshop with wood and tools in it. These areas were then decorated with photographs and posters of women involved in some form of construction work.[3]

The contents of this chapter are the result of a number of years' work over which time our thoughts and ideas have developed and changed. Thus our description is of an ongoing process rather than a definitive statement.

Notes

1 Our enthusiasm and interest was encouraged and supported by Hazel Taylor, who was then our Equal Opportunities Adviser, and Fiona Collins, who had been seconded to support equal opportunities initiatives in primary schools.
2 Hazel Taylor in an article in *Girls into Maths Can Go* edited by L. Burton.
3 The photographs and posters came from the 'Working Now' pack produced by Brent and now Birmingham Development Education Centre, also Women in Engineering and Women in Construction (ILEA).

CHAPTER 8

Challenging Sexism – Cycling Yesterday and Today: An Account of a Cross-Curricular Project

JULIE CAHILL and UMA PANDYA

Introduction

This chapter provides an account of a cross-curricular project which was developed in a middle-infant class with 5- and 6-year-olds. It lasted for more than half a term and the planning and co-ordination was undertaken by two teachers within the school and the Craft Design Technology (CDT) Primary Advisory Teacher.

The project – bicycles and wheels

The project was precipitated by three factors:

1 A child's bicycle was brought into school by a parent. It needed mending before it could be used.
2 A story book, *The Ten Woman Bicycle* raised issues about women and technology.
3 The potential of developing such a project as a good example of anti-racist and anti-sexist work which could form the basis for discussion with other teachers as part of the CDT in-service training.

We decided to make a video; something we had never attempted before. It thus provided us with a good opportunity to acquire new skills. The video was made by teachers involved in the project after some initial assistance from a woman teacher involved in a community video production course. The teachers felt that it was important to retain positive female role models in this and all aspects of the project. The children could see women operating videocameras and have first-hand experience of working with women engaged in non-traditional, technical roles. In addition, the children could see the camera in action and also realize that a high value was placed on their work which we wished to film.

The project was developed in Mora Infant School which is situated in Cricklewood, in north-west London. The school is part of a multiracial community and the school aims to develop in children, boys and girls together, an awareness of their own potential. The teachers encourage critical thought, investigative learning and the development of children's ideas. The children are also encouraged to represent their work in a variety of ways, including written and spoken forms. The teachers also believe in the importance of children working together as this enables them to learn to respect and value each other and to feel an important member of the school community. Over the past few years the staff, parents and governors have actively supported the school's anti-racist and anti-sexist policies.

The teachers' planning of the curriculum, their classroom practice and the ethos of the school are all aimed at enabling each child to develop their full potential. When planning any topic, teachers need to ask themselves not only whether it meets the requirements of the National Curriculum but also, according to LEA policy, whether it is 'broad, balanced, meaningful and holistic' (Brent LEA, 1987). It is important that topics fulfil these criteria as a curriculum which addresses these through the process of learning as well as through skills and knowledge taught will help redress inequality.

Drawing on learners' own experiences makes the curriculum more meaningful and is more likely to ensure commitment. At various levels this has meant that the school has taken positive action; for example, involving girls in construction activities and 'tinkering', with the joint aim of developing girls' skills and knowledge and a confident approach to the work. The teachers try to ensure that children are presented with positive role models in terms of gender, race and class; for example, parents, visitors and women teachers are involved in non-stereotypical activities and resources are used that challenge stereotypes and promote discussion.

The staff believe that all children should have equal access to all areas of the curriculum in a way that develops positive attitudes towards themselves and each other. In so doing the children are provided with choices that will enable them to determine the direction they wish to take and develop in their lives.

Curriculum areas covered by the project

The organization of the primary school curriculum is conducive to the incorporation of designing and making activities. Teachers in the school usually plan the term ahead around particular concepts, which allows flexibility in terms of integrating various areas of learning and experience. The inclusion of CDT activities in this context enhances the existing curriculum. The school operates an integrated day and great emphasis is placed on collaborative and co-operative work. This was essential to the project.

Designing and making

Children closely observed the working mechanics of real bicycles. Conversations with girls highlighted the depth of their understanding. For example, a group of children and a woman mechanic started to dismantle parts from a real bicycle and when one wheel was taken off, the following conversation took place.

> *Mechanic*: Will the other one come off?
> *Girl 1*:   Only if the bar can come out.
> Two girls and a boy in the group then retorted, 'No, the other bar can't come out'.

Working on real bicycles presented the opportunity for children to work collaboratively with the girls sharing the knowledge and skills equally with the boys. At one point two children (a girl and a boy) were cleaning the rust off the bicycle with a girl turning the pedals. During the process many mechanical ideas were introduced, for example, the concept of lubricating parts for effective use. When children grasped this principle, they automatically lubricated the other wheel and, having cleaned the chain, one of the girls said, 'We need some more oil here'. The teacher working with another group on cleaning and restoring the brakes asked, 'If you are riding along on this bike and suddenly need to stop what could you use to stop?' The unanimous answer was 'The brakes!'

Teacher: Has this bike got brakes?
Girl 1: (looking at brake lever) Yes.
Girl 2: (looking at the wheel) No.

A discussion ensued clarifying the distinction between a brake and a brake lever. After this a girl in the group commented, of her own accord, 'If you don't have brakes, you can't stop. If you don't have pedals you can't ride it'.

These comments illustrate the technical understanding that the children were developing as a result of the project.

'Tinkering' is actively encouraged throughout the school. This has occurred at times in an all-girls group. The need for this arose when observation of mixed groups showed that boys tended to dominate and monopolize kits and tools. The children's experience of 'tinkering' meant that they experienced few problems in handling and manipulating the tools required in relation to the project. Familiarizing children with the correct names of tools (e.g. 'Can you get the screwdriver?') increased the children's confidence. This project broadened and extended the children's experience of using different tools within a meaningful context.

After spending some time looking closely at wheels the children decided to make their own using various materials and real tools. The teacher asked a girl what sort of wheels she was making.

Girl: A wheel for a bike.
Teacher: What do wheels on bikes have?
Girl: They have spokes.

Another girl in the group wrote a vivid description of the process of making a wheel:

We got a cardboard circle. Then I got some sticks.
First I had to measure the spokes. After that I used
a clamp and a hacksaw to cut the right size spokes.
Then we put the spokes into the wheel.

These two examples show that close observation of the working mechanics of real bicycles had equipped children with knowledge which they translated into practical terms by designing and making their own models. This was carried out using a variety of media as well as using different technical skills for creating joints and movement. The models made with construction kits were tried, tested and modified where necessary. After making a model of a bike from the Big Builder, one girl commented 'My bike looks like the nursery's'. Two other girls showed the teacher their completed models of bicycles saying 'They balance!' The children recorded their work through writing, drawing and painting with intricate labelling of parts.

Many of the skills, knowledge and values covered by the project were also mentioned in the National Curriculum Design and Technology Report (June, 1989).

Language

Use of slides depicting bicycles in everyday life around the world invoked memories for some children which they were able to share with the class. The slides also stimulated discussion on the appropriateness of vehicles in various environments. The school's positive attitude towards multilingualism enabled children to respond in whatever language they felt most comfortable with at the time. For example, the presence of Aisha, a Black woman mechanic who spoke Urdu, enabled the Urdu-speaking children, and in particular the girls, to ask technical questions in their home language. This is something they would have been unable to do had only English been used as the medium of communication.

Discussion was also stimulated by stories, including stories told by Wilf, the assistant caretaker, of his cycle-racing days in the Caribbean. All the designing and making activities, as well as work in other curriculum areas, involved group and class discussion.

When a group of children were involved in finger-painting wheels, one child mentioned seeing a wheelchair. The teacher took the opportunity to ask questions, sensitively, about wheelchairs and their use and, in so doing, helped the children to explore some of the issues related to disability.

The children made class and individual books. In one book, the wheels on each page were attached with paper fasteners so that the wheels turned. The teacher showed the children how to do this after the children had looked at several books with moving parts. The children decided what to make and how they could build pictures around moving wheels. Photographs of the children and their work were also used in bookmaking. This raised the children's self-esteem by giving status and value to their work and provided a lasting record of their achievements.

Computer

Much of the children's writing was done on a BBC computer using the folio writing program. The computer had been in the classroom throughout the project and the children were already confident about working with it. Printed work included factual writing (e.g. the process of making a model) and creative stories. Pattern and design programs were used with other curriculum areas.

Art

A wide range of materials were used to encourage and stimulate the children's creativity. These included paint, pastels, chalk, pencils, junk and scrap bank materials as well as many others. The children worked collectively and individually. Processes and skills included printing, painting, rubbings, still-life (observational drawing), model making with pliable and rigid materials, finger painting, patterning, experimenting and using tools. Children's cultural values were evident in much of their artwork from models to patterns. This was encouraged by the global perspective of the project and the status given to children's work. A large frieze was made of *The Ten Woman Bicycle* story which showed women with strong, positive images. This encouraged some girls to learn to ride bikes and provided a focus for discussion around the story. Again, there were strong links with other curriculum areas, particularly CDT and maths.

Mathematics

Mathematical work flowed easily from most of the work during the project. In this class there were no apparent gender differences in relation to maths because of strategies used by the teacher and the nature of the class composition (the girls were particularly assertive). The class had been taught by the same teacher the previous year, during which time the children had experienced a predominantly practical approach to maths in the context of cross-curricular topic work and there had been an emphasis on the development of spatial concepts.

Bicycles and tricycles led naturally to discussion about the meaning of bi- and tri- and then on to counting in twos and threes. Because of the focus on wheels, circles became significant to the children and they noticed them everywhere. Size and measuring plus planning were important for lots of activities. As regards language, children felt able to count in both English and Urdu (even non-Urdu speakers). Within the school, children were encouraged to use their first language and much of the work displayed was labelled in Gujerati, Urdu and English. Because of the status given to languages other than English, other children are often keen to join in with bilingual children.

The project was carried out before the introduction of the National Curriculum but when we matched the project with the Maths Attainment Targets we found that the mathematical skills and knowledge developed during the project related to Attainment Targets 1, 2, 3, 9 and 10 at the appropriate level in the National Curriculum.

Science

Specific scientific skills and knowledge were developed by this project. The children were involved in close observation, asking questions, hypothesizing, devising possible solutions to problems, testing and evaluating their own suggestions and modifying their ideas in the light of their experiences. There were clear links between science and technology. When, for example, the children were engaged in building models they were also exploring the properties and uses of different materials. The project also introduced a wide range of scientific ideas in a very practical context. The understandings that the children developed of such concepts as the effect of pushes and pulls on the movement of things, friction, the operation of gears, the cause and effect of rust and causes of environmental pollution will be built on later in their school careers.

Visits

The children visited the Commonwealth Institute where a tricycle rickshaw caused great excitement, recognition (particularly by children whose families were from Pakistan and India) and discussion. This was all later reflected in the children's work. Later on, at the end of the project, the children visited the science museum. Anything with wheels or gears became an immediate focus of attention. The children had not lost interest in the project.

Parental and community involvement

The project was something that all children and parents could relate to and we actively sought the involvement of adults within the school, other than teachers. As a result, enthusiasm quickly developed and links between home and school ensured good working relationships.

The school

Other children in the school also developed an interest in the project for the reasons that parents and other members of the community had become involved. All the staff helped us to write new words for a song which was familiar to the children. It became so popular that the whole school learned it and it was performed complete with movements and musical instruments during the school's winter assembly. Through this other children became indirectly involved in the project.

Design and making and the National Curriculum

At the time of writing, many teachers are anxious about the implications of the National Curriculum. We are optimistic as we believe that the main changes brought about by the National Curriculum will be concerned with assessment whilst curriculum content and delivery will still be left to schools and teacher's judgements and creativity.

We were heartened by the The Interim Report of the National Curriculum Design and Technology Working Group (November, 1989). The group placed great emphasis on 'breadth, balance and relevance'. Earlier in this chapter we stated that these are some of the things teachers should take into account when planning

their curriculum. The Interim Report also highlighted the importance of the context within which design and technology activities
occur:

> We believe it is important that Design and Technology activities
> should be undertaken by girls and boys in a range of contexts.

Selecting contexts within which to place and undertake the work
is crucial if children's motivation to learn and their interest is to
be sustained. Of course, when selecting contexts, it is imperative
that the context of design and making is not 'off-putting' to girls
but is of interest to them and, in addition, will motivate, draw
on and value the experiences of both Black and White children.
Our project, which looked at the uses of bicycles around the world,
stimulated discussions about rickshaws (bicycle taxis used in India
and Pakistan). Some of the children in the class had been to India
and Pakistan and therefore could relate their experiences and
discuss with enormous confidence and enthusiasm. Seeing real
rickshaws at the Commonwealth Institute was well worth the visit.

'Values' were also used as a starting point in the project and
these were also prominent in the 1989 Interim Report (pp. 18–19).
Through the project we aimed to challenge stereotypes (e.g. having
a Black Asian woman repair a bicycle must have gone some way
towards challenging stereotypes of the roles of Asian women).
Environmental issues raised included pollution, safety of children
coming to school by bike with their parents and those who came
by car and the traffic and safety problems near the school caused
by parents in cars in the morning and after school.

Conclusion

We felt that this project was a good example of equal opportunities
work in the early years of schooling as it fulfilled the criteria
referred to earlier. The project was 'broad, balanced meaningful
and appropriate'. The project shows how it is possible to select
a learning context that is relevant to the experiences of all the
children in the class, girls and boys and also girls and boys from
a range of cultural backgrounds. The project was broad and
balanced as it encompassed a wide range of knowledge and skills
from different curriculum areas. It was meaningful and appropriate because it drew on the children's own experiences and

involved their parents, other members of the community and some members of the school community other than teachers. This project fulfilled many of the requirements of the maths, science and design and technology elements of the National Curriculum although it took place before the National Curriculum was introduced.

We feel that this project gives an indication of how schools can implement both the National Curriculum and their Equal Opportunities Policies.

Notes

At a national conference in 1988, this project was awarded 1st place by the Fawcett Society for its contribution to equal opportunities. This was a positive reward for the school and the borough for their policies on equal opportunities.

The authors would like to thank Myra Joyce, Deputy Headteacher of Mora Infant School, whose class was involved in this project.

CHAPTER 9

Planning and Assessment in the Early Years

JANE SAVAGE

Assessment is often seen as the final action to be undertaken by the teacher at the end of an activity or project. However, to be a useful tool for a busy early-years teacher it is necessary to view assessment as part of the planning process that enables individual children to build on their previous experiences and to develop both individual skills and more generalized educational goals. To be effective at modifying teaching strategies, assessment has to take account of what is to be taught as well as how it is to be taught and learnt.

Assessment is an activity that involves teachers, children and parents although there is often no structure in existence that will involve all these parties. Methods such as those involved in the ILEA's Primary Language Record are pioneering the idea of shared recording.

There are two main forms of assessment; the summative report type which is usually undertaken at the end of the academic year and the more diagnostic cumulative assessment which takes place in some form whenever children and adults interact. Both need to be examined to determine how they can become successful teaching and learning tools to ensure equal opportunities and relevant learning activities for children as well as being useful tools for the classroom teacher.

In order for teachers to be fully aware of the progress and development of the children in their class it is necessary for them to have an understanding of what skills, attitudes and ideas the children have already encountered and what it is intended they will cover. You can only ascertain how successful a teaching strategy has been if you have started from a clear set of aims and objectives. There is a further complication when we consider assessing early-years science and technology work and that is that many early-years teachers have had limited science experiences and do not feel confident to teach either science content or scientific skills and processes.

It is important to develop workable assessment strategies that will assist teachers to monitor the progress of the children in their care, are sophisticated enough to focus on children's individual special needs and address key principles such as equal access to the curriculum. In order to develop such assessment strategies it will be useful to consider the following factors:

1 How do boys and girls differ in terms of their involvement in the curriculum?

2 How do other factors such as race, class and gender affect children's access to the curriculum?

3 How are young children given opportunities to work in a variety of ways?

4 How are specific science and technology activities structured and what is their purpose?

5 How do teacher expectations and perceptions of primary science and the children affect the organization of the classroom, the activities that are available and who is able to do the activities?

6 How are teachers, children and parents involved in assessing the children's work, activities and progress?

7 How can the practical problems of assessing approximately 30 individual children be achieved by one class teacher in a busy early-years classroom?

8 How can teachers provide continuity and progression within their classrooms and in their schools?

9 How can assessment become a positive diagnostic tool that is of use to all involved in the learning process?

There have been many studies that have highlighted the different opportunities that are open to children in primary schools. It is not good enough for teachers to assume that all the children in their classes have equal access to the science curriculum. Teachers must develop mechanisms to monitor which children are involved in different activities, pinpoint who is slipping through the net when it comes to certain activities and from here develop strategies that ensure that all children have the chance to use particular pieces of equipment or work in a certain way. Children must be given the opportunities to work in many different ways so that the teacher can make informed decisions about which are the best ways for an individual child. Teachers need to monitor which groups of children are working together, whether a child has had an opportunity to develop skills individually, as a paired activity, as a small group, as a large group and as a class. If the teacher becomes aware of these different ways of working then she can plan work which provides both equal access and a variety of working groups for all children. This will also enable teachers to identify which working group arrangements are most suited to individual children and which kind of activities a child needs support in. For example, some children rarely take an active role in any group of more than six children. Some children work in a way that encourages an open-ended problem-solving approach when they are with peers of the same sex, sometimes a mixed sex group is best.

A tick sheet like the one below is a quick and easy way of enabling teachers to become more aware of how children are operating in their classroom.

Activity:

• Children who worked as individuals:
• Children who worked as pairs:
• Children who worked as a small group [up to 4]:
• Children who worked as a larger group [4 to 10]:
• Children who worked as a whole class:
• Children who have worked in a mixed-sex group:
• Children who have worked in a single-sex group.

Science has traditionally been a subject with which many girls have felt 'uncomfortable'. Girls have often not had equal access to science and have been constrained by peer group pressure, the nature of the science that has been taught and the behaviour of some children who can dominate a group, so that opportunities in science either passed them by or seemed unattractive. These are some of the background reasons why early-years and primary teachers, the majority of whom are women, feel that they lack expertise or the confidence to teach science. In fact much science work has always been incorporated into 'good early-years practice'. Teachers must reflect on their own practice to realize what they are and are not providing.

When assessing a child, teachers need to be aware of how the child's performance may be influenced by the activity the child is engaged in, the resources and the language used. For example, some practical explorations which are used in published schemes of work are based on activities such as mowing a lawn, having special clothing for rainy weather (e.g. wellingtons and macs) or cooking fudge and toffee. Many children have none of these experiences and would be placed at a disadvantage if teachers assumed that the understanding and language resulting from such experiences were common to all children. It would be better if activities involved experiences shared by all children or those where similarities between different practices can be explored (e.g. what happens to our clothes when it rains? How do we use sugar to make celebration sweets?). Children are also unlikely to operate at the peak of their potential if the activity is of no interest to them.

Teachers also need to think about the language used during assessment. It is important that gender-neutral language is used and that teachers do not assume that all children understand the meaning of seemingly 'common' phrases or words. When assessing bilingual children, teachers may need to consider translating keywords or phrases. If, when assessing children, we present them with activities with which they are unfamiliar, or activities where the language used is unfamiliar we place children at a disadvantage as we are not enabling the children to utilize their previous experiences and knowledge. Furthermore, if teachers only assess in terms of a certain type of limited experience then all they may really be assessing is who has experienced this activity before. Such an assessment will not provide the teacher with accurate, positive information about what individual children can do.

It is important to check that resources which are being used are non-racist and present positive, active, cultural role models. Even some of the most recently published science materials for use with young children present images which carry both explicit and implicit messages about who 'should' be doing an activity.

End-of-year reports in which teachers attempt to sum up a child's achievements and progress for an academic year have long been a source of dissatisfaction to early-years teachers. Although many versions of these end-of-year reports exist many teachers have found them limited and, because there have been so many important curriculum developments in recent years, out of date. Some report sheets which are in use today do not even have a section in which to record primary science activities.

The ILEA's Primary Language Record or PLR is an attempt to improve many of the ideas associated with summative and cumulative record keeping. It starts from the premise that assessment involves gathering information from teachers, all those who work with a child, the child itself and most importantly, its parents. An evaluation is then made which is based on these shared observations and opinions. Various parts of the PLR record sheets are filled in throughout the academic year, for example, parental interviews are usually done in the autumn term. An important part of the PLR is that parents and children to contribute to the record so that achievements and concerns can be shared and recorded in an honest and open way. Although the PLR is concerned with assessing and evaluating language development many of the strategies could be used successfully for other areas of the curriculum including science and technology.

The practical problems of assessing a large number of young children are enormous but not insurmountable. If assessment is seen as an integral part of planning then mechanisms can be built in which allow the teacher to monitor and evaluate the progress of individual children. The sampling of children's work is a technique that can help teachers to assess a child's progress and development. Sampling children's work is a useful way of providing concrete evidence of a child's work and how it has changed over time. However, there are many factors which need to be considered and ideally discussed by staff if sampling is to provide a useful aid to teacher's observations about individual children. Teachers need to consider:

- When samples should be taken and how frequently
- If actual pieces of work should be sampled or if work should be photocopied
- What types of work should be sampled
- How should large and three-dimensional pieces of work be sampled; are photographs to be taken?
- If the piece of work to be sampled should be of average quality, or a best or breakthrough piece.

In this way teachers will be able to build up a clearer picture of on what it is they want to base their observations and how these observations can help them to evaluate a child's needs.

Organizational 'tick sheets' such as the one on working methods outlined earlier in this chapter can also help to monitor how individual children are working. Another useful strategy is to focus on one child each day, or a small group of children each week, or a particular skill, attitude or process. Checklists are a quick way of getting a limited amount of information. There is much that they cannot tell you. Child interviews where you discuss work and ideas with individual children are much more revealing and detailed but are also more time consuming and difficult to organize. Discussion is vital in order to get a clear picture of work that has gone on which the teacher may not have been closely involved in. In this way a teacher is setting his or herself more easily achievable goals in terms of record keeping and providing themselves with accurate points based on observations.

When observing and discussing children's work it is useful to have a list of skills, processes and attitudes to focus observations. This list could be based on Attainment Target 1 and its associated guidance in the Science National Curriculum. Other useful checklists can be found in Wynne Harlen's work (Harlen, 1977, 1985).

All teachers should develop their planning, recording and assessment skills and strategies in order to deliver the curriculum and to help each child to progress and develop at his or her own pace.

Planning and assessment skills such as observing, discussing, making regular, focused records and developing planning strategies are important because whatever systems are introduced these skills will play an important part in them.

One of the major advantages that has come about through the introduction of the National Curriculum is the fact that teachers now have to plan on a longer timescale than they have previously been used to. To deliver a balanced curriculum throughout the academic year teachers have to plan for the year. In addition in order to provide continuity and progression throughout the school, teachers have to share both their planning and their evaluations of what was actually achieved. Together they should come to a clearer, shared understanding and analysis of what children have already encountered and what they should go on to meet.

For assessment to play an active part in the learning process of the children we work with it must be a useful tool for the teacher. It should be closely linked with planning; in order to provide equal opportunities and status for all children involved in early-years science work we will need to highlight what is being delivered, how it is being supported with resources and how each child is involved in it. Many everyday activities have a scientific or a technological component to them and it is necessary to be aware of these and to base our science work with young children on them. All children and their teachers are scientists. Careful planning and evaluation of those plans will allow children and teachers to build on their individual science skills and attitudes.

CHAPTER 10

Inquiring Girls: An Examination of Published Science Education Materials

PATIENCE MACGREGOR and CLAUDETTE WILLIAMS

Until recently teachers in the early years (3–8 years) have tended not to be explicit about the science curriculum they provided. Generally, they have taught what they felt was appropriate for this age, without consciously identifying a physical science content in what they taught. Work therefore in the early years tended to centre around the biological aspect of science, for example, growth, nurturing and care. In part, the experience and expectations of women teachers, who predominate in the early-years phase, gave rise to this tendency to focus on the biological aspect of science education, because this was the nature of the science they received during their own schooling.

In addition it could be said that staff 'locked' into 'child-centredness' (Curtis, 1986) to justify this limited model of science education (See Chapter 1). Science teaching was identified by HMIs as involving specific skills learned through the processes of observation, classification, hypothesis, experimentation, recording and evaluation (DES, 1983). A conflict arises between child-centred education and the need for a more structured approach in terms of the practical, explorative nature of physical scientific investigation. In the early-years phase, where there is a tendency for the focus to be on skills, teachers often appear unsure as to how to extend the learning experiences in relation to the scientific processes.

The increased emphasis on teacher accountability and the need for teachers to communicate what they do and what they actually know about children's learning, has given rise to an increasing number of local education authority documents and textbooks defining an early-years science curriculum. The National Curriculum further supports teachers by bridging the gap between skills and processes and by locating the learning within an integrated curriculum (topic/thematic approach).

To those early-years staff who lack confidence in their own knowledge of science, meeting the demands of the National Curriculum may seem awesome, however, there is extensive published resource material in the form of books, with ideas for teachers on science content. It is, however, more difficult to find material on the theory and methodology of the teaching of science in the early years. Staff will have to develop strategies to help children become confident, competent science learners and, in particular, to develop strategies which are going to support girls towards this end.

Selecting material

We have taken a critical look at some of the published materials available to early-years staff, either as a teacher resource or as a child resource. We found it imperative to be clear about the criteria we would use to select materials and the following check list outlines the questions we asked ourselves.

- Is the material permeated with an equal opportunities perspective in the context of race, class, gender and disability? – materials which were consistent in challenging stereotypical attitudes of people, and in presenting alternative views of the world.
- Is the scientific content stimulating, interesting and challenging? – material which was open-ended and structured in such a way as to motivate children to solve problems themselves.
- Does it extend, develop and reinforce concepts and ideas and is it suitable for children at different stages of development? – material which involved practical activity so that children could hypothesize and test what they know, and move on into new areas of knowledge.

- Are there a variety of starting points and are there a range of styles of learning possible? – material which could be used with different class groupings, where knowledge could be shared and collaborative learning was encouraged as well as independent learning when appropriate.
- Is the material relevant and accessible to all children, and does it build on children's experiences and introduce new ideas? – material which represented a wide range of peoples and communities in a variety of situations.
- Is there opportunity for parental and community involvement? – materials which gave opportunity for parents and members of the local community's skills and knowledge to be brought into the school/class topic.
- Is the text appropriate for the scientific ideas and concepts, within the context of the children's cognitive development? – material which matched language and task at the appropriate learning stage of children in the early years, for example, we excluded material which 'talked down' or 'talked over the heads' of children.

When we looked at materials and applied our criteria, we found that no single set of books for teachers, or resource packs for children complied with all the criteria. Early-years workers will need to select from a variety of sources, and will need to continually review and update their resources.

We found that, generally, little account was taken of gender and race issues, but that there had been a significant change during the last six or seven years when publishers had begun to make token, even if inadequate, gestures. For example, in Ginn Science Happening, Aylesbury (1970) all illustrations were of White characters and were sexist but Ginn Science (1988/1989) included a few Black children.

Not only did earlier publications take little account of gender and race issues, but they also perpetuated the view that science is 'neutral' and a 'Western European' creation (Gill and Levidow, 1987). By this we mean the resources fail to recognize, and consciously exclude science originating in other societies and cultures which have influenced and enriched 'Western' scientific knowledge. We would argue that this failure deprives children, functioning in a multicultural community, from fully appreciating other views of the world. Likewise children who have family connections in other parts of the world do not get this experience confirmed and validated.

All the publications we examined failed on the above points. In this context we recommend that early-years staff become familiar with an enterprising series *Third World Science Project* (William, 1982). The ideas and information incorporated in the units come mainly from science teachers who are, or have been, teaching abroad. In some cases their pupils have also contributed. While this series is aimed at older children, the topics have a lot of scope for development within the early-years phase. Ivan Van Sertima's book *Blacks in Science* (1988) and D. M. Salwi's *Our Scientists* (1986) provide an African and South Asian perspective which challenge existing, taken for granted, scientific practices.

Review of published resources

School Council 5–13, Macdonald, 1972–1973.

A complete set of 27 titles. It is aimed at the 5–13-year age range but many of the ideas covered can be adapted for 3–5-year-olds. These books are a teacher's resource, full of good ideas, but not 'user-friendly', unless staff have plenty of time to become familiar with them. We found the guidelines at the back useful and worth keeping in mind. Teachers will need to adapt some of the material and ideas to expand children's knowledge of the world.

Science in a Topic, Hutton, 1977–1981.

The complete set has nine topic books and a teacher's guide. A useful set of teachers' resources for exploring ways of working through a topic across the curriculum. We would not offer them directly to children as the images certainly need to be questioned in terms of an equal opportunity perspective.

Avon Primary Science Working Papers, Maths, Science and Technology Centre, 1985–1989.

The set contains seven working papers and a primary science guideline manual (revised in 1989 to meet the needs of the National Curriculum). They are extremely useful and cover most aspects of early-years science, but are not consistent on issues of equal opportunity. They are topic based, containing good case studies and are written in conjunction with practising teachers. The ideas are open-ended with many starting points, thus making it possible for teachers to include an equal opportunities perspective.

Science for Children with Learning Difficulties,
Macdonald, 1983.

A unit for teachers, with clear ideas and practical investigations, followed by some discussion points which help bring out scientific ideas.

Into Science by Terry Jennings, Oxford University Press 1988.

The six books in the series are well illustrated, with familiar topics deriving from the themes of change, living things, forces and energy, structures and materials. Each book includes notes for teachers and parents and a glossary of terms.

An Early Start to Science, Macdonald Education, 1987.

This is an activity book, full of ideas, which addresses gender and race. The layout suggests that it is a teacher's resource rather than a resource for children. The text and illustrations are not always clearly matched and it is ambiguous as to who the reader might be. The layout is densely packed making reading difficult. In addition, an idea is sometimes just touched on whilst at other times, ideas are explored in far greater depth.

Third World Science by: Professor Iolo Wyn Williams,
Third World Science Project, 1982.

Third World Science project seeks to develop an appreciation of:

1 The boundless fascination of the natural world
2 The knowledge, skills and expertise possessed by men and women everywhere
3 The application of knowledge, skills and expertise to solve the practical problems of everyday life
4 The impact of modern technology in the Third World
5 The influence of the cultural background on the perception of knowledge, problems and solutions

The following units are available: Carrying Loads on Heads; Charcoal; Clay Pots; Dental Care; Distillation; Energy Convertors; Fermentation; Housing; Iron Smelting; Methane Digestors; Natural Dyes; Plants and Medicines; Salt and Soap.

Suggestions and feedbacks are invited from users concerning the use made of the material, pupils' reactions, suggestions for any additions or improvements.

Look! Project, Oliver and Boyd, 1984.

Look! Primary Science ages 4 to 8, views science as a process, a way of understanding the world, in a social, cultural and technologically significant way. The child, home and family, the natural and made environment, materials and energy are the main areas it focuses on. Issues of equal opportunity need closer attention and the quality of the illustrations is poor. The Look! Guide to Primary Technology Policy and Guide to Primary Science Policy provide excellent systematic approaches to help schools plan their individual school policies, with clear advice on organization and content.

Ginn Science, Ginn and Co. Ltd, 1989

Ginn has a strong emphasis on skills – 'the way in which scientists work'. It is clearly laid out into levels 1 to 3, with each level having a class kit, a starter pack, a resource file, a group discussion book, and story books. Level 3 has accompanying assessment booklets (Figure 10.1). There is an emphasis on oral work and group discussion, with photo books to stimulate the children. This makes this scheme very attractive to early-years practitioners. The 'science' though, is presented within a very narrow world view, and issues of equal opportunity are hardly addressed.

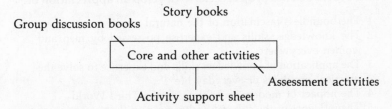

Figure 10.1 Structure of scheme

Longman Scienceworld edited by Brenda Prestt, Longman, 1988.

Longman Scienceworld uses familiar science topics such as, clothes, spring and the school caretaker as starting points. It has an integrated approach to infant and junior science topics, and the material is divided by the age groupings 5 to 7 (infant) and 7 to 11 (junior).

The scheme contains three starter books for infants, which are large fliptype books, to be used to stimulate group discussions. The three teachers' books that go with these are full of practical activities, with step-by-step instruction for those not very familiar with working with a thematic approach. The illustrations present a variety of images of children and adults working in different contexts. We would not recommend the starter readers, as the language used is stilted and not very meaningful.

Helping Children to Become Scientific, ILEA Guidelines for Teachers. Centre for Learning Resources, 1988.

As a guideline we strongly recommend this book and found it to be excellent value for money. It addresses the issues raised by our criteria, in a user-friendly format. It includes for example, classroom organization, planning, assessment, record keeping and methods of organizing a whole-school policy towards science. The black and white photographs used to illustrate children's activity are visually attractive and strengthen the text. There is a useful bibliography at the back. The book discusses the scientific processes and develops ideas through topics with a sense of progression built in, and is therefore most useful in supporting staff with the National Curriculum specifications.

We also looked briefly at the Schools Television Broadcasts relevant to science matters for the age range 3–8 years. What we saw was extremely disappointing. During our two weeks viewing, approximately 80 per cent of the science programmes we saw referred to the biological science rather than the physical science.

Moreover, women as science makers were invisible. The programmes essentially were about White men, made by men and fronted or narrated by White men.

We felt that the programmes offered very little scientific challenge and could not be said to utilize the creative and imaginative power or possibilities of the television medium in developing ideas. Many children we know watch 'Tomorrow's World', a programme broadcast at prime viewing time, and not necessarily aimed at young children, which uses a far more exciting and informative presentation as well as looking at the physical sciences in the main. The programme is presented by two able, competent unstereotypical women.

Conclusions

In summary, in our examination of some of the science resources available to staff and children in the early years, we have been encouraged to find that there is a wealth of materials supporting scientific learning. However, we have found that no one scheme met all our criteria and in fact some of the schemes gave cause for concern on issues of equal opportunities. Publishers and authors need to take on their responsibilities and create learning materials which reflect the issues of equal opportunity in a pluralist society. Until they do, staff will need to be selective in terms of choosing resources and combining schemes to meet the needs of children, and critical and demanding of authors and publishers. We found the *Avon Primary Science Working Papers* and the ILEA's guidelines, *Helping Children to Become Scientific*, most useful in initiating thought and discussion about implementing good early-years science practice, as well as building on teachers' existing scientific skills and knowledge.

We felt a need in some of the schemes for some clear explanation of how the individual parts related to each other, with more indication of how to develop them in broader terms, rather than the simple progression from level 1 to level 2 or from Infant to Junior.

The more recently published schemes and books have begun to address equal opportunity issues, but the whole concept needs to be widened and continually addressed. As there is a need to select from a variety of materials, we felt there is a need to seek out supporting resources which present a wider world view. These resources can be found from sources such as parents, books and posters published in other countries, artefacts and materials from agencies working with other nations, for example, Oxfam, Cafod, the ANC and Traidcraft, etc.

The dissemination of information and sharing of resources can be the most positive way forward for staff in the early years, leading towards a shared policy on science, which acknowledges equal opportunities and views science as meaningful to every child.

CHAPTER 11

Science INSET and Equal Opportunities in the Early Years

BARBARA WYVILL

Draw a scientist

'Draw a picture of a scientist doing something'. For three minutes 18 nursery and primary teachers followed the instruction with a few embarrassed smiles at the thought of having to practise a little-used skill. My role was simply to encourage their creative efforts. This was the introductory activity of an In-Service Education of Teachers (INSET) session on equal opportunities in science in the primary classroom held at the ILEA Gordon Teachers' Centre as part of ILEA's Division 6 Women's Week (March 1989). I had two purposes in setting this activity. Firstly, to show that most people's idea of a scientist is a stereotyped one and secondly, to elicit some ideas on the meaning of 'science' by looking at it as an activity which is carried out by scientists.

A scientist stereotyped

The drawings were each awarded 25 marks for artistic ability. Marks were then deducted for portraying scientists as male, middle-class, old (over 21), wearing a white coat and eccentric. Those who had drawn a scientist with all five of these attributes lost all of the 25 marks.

This exercise has been repeated with teachers and children and, in general, the scores have been very low showing that most of the people tested held a stereotypical view of scientists as old, eccentric, middle-class males in white coats. This is hardly surprising since the scientists that they see in the media or read about in history books nearly always conform to this stereotype. But there are and have been many female scientists and Black and Asian scientists with various social backgrounds. Children too, as scientists, can perform investigations and experiments to explore the world around them.

Science is something a scientist does

The second purpose of the 'draw a scientist' exercise was to explore ideas about the meaning of science. I asked each teacher, in turn, to say what her or his scientist was doing and wrote down a list of the scientific skills which the scientists were using, and also the fields of investigation or study.

Scientific skills	Field of investigation
Observing	Plants
Finding out	Chemistry
Exploring	The world about us
Investigating	The stars
Thinking	Bubbles
Asking questions	Rocks
Making hypotheses	Medicine
Predicting	Flowers
Measuring	Kites
Testing	Transport
Experimenting	Windpower
Recording	Electricity
Analysing	Animal behaviour
Problem solving	Buildings
Discovering	Water
Communicating	Ourselves

I then asked the groups if these lists were a complete summary of their ideas about science or whether there were other aspects not covered by the two lists. The consensus was that science also has political, moral, economic and social aspects and that people could learn social and manipulative skills through scientific activities.

Having thus discussed the meaning of science while avoiding the difficult task of defining it, we then returned to the question: is there something in the nature of science itself that prevents it from being accessible to women? The list of scientific skills does not contain any skill that a woman cannot acquire equally as well as a man. The fields of study and investigations covered by science included the whole of the physical, biological and constructed environment – all areas that equally affect the lives of women and men and people of different ethnic origins and social class. People of both genders from all ethnic groups and social classes are also capable of carrying out scientific activities.

Scientists and socialization

Relatively few scientists in Britain are women or Black or working class (see Table 11.1). Why is it that so many female primary and nursery teachers also lack confidence and competence in science. I would argue that there is nothing in the nature of science that prevents it from being accessible to women. The fact that there are successful female scientists and science teachers and women who can use scientific skills in their daily lives shows that women have the capability to become scientific. Perhaps the answer lies in the different ways society socializes women and men.

Table 11.1 Number of men and women employed in selected occupations – GB 1981

	Thousands	
	Males	*Females*
Professional occupations		
Medical practitioners	60	19
Economists, statisticians, System analysts, programmers	78	19
Teachers in higher education	84	30
Scientists, engineers, technologists	896	87
Other teachers	239	539
Nurses and nurse administrators	48	539
Non-professional occupations		
Domestic staff and school helpers	10	521
Secretaries, typists, receptionists	17	861
Total in employment	13 765	9 151

(Source: Census 1981: Economic Activity, Great Britain)

The Girls Into Science and Technology (GIST) Project found that at the age of 11, while scoring equally on tests of scientific knowledge, boys and girls have already developed strongly differentiated attitudes towards science – with boys being far more interested in the physical sciences (Whyte, 1986: 17–21). These differences I suggest are not innate. The process begins when infants are treated differently according to their gender. There is considerable evidence to show that gender-stereotyping begins at birth and by the age of 1 there are noticeable differences between the play of girls and boys (Goldberg and Lewis, 1969: 21–31). Manufacturers and retailers often divide toys into girls' toys and boys' toys. Manufacturers can label toys quite subtly. By carefully selecting images of girls and boys on the packaging of toys they can show exactly for whom they are designed. For example, the cardboard lids of chemistry sets and construction kits are decorated, mainly, with pictures of White boys. These are just a few of the influences on the development of girls and boys.

Raising awareness

It is possible to encourage teachers to examine their own socialization and experiences by playing the following game. The teachers stand in a row across the middle of a large room facing the front. They are then asked a series of questions. If the answer to a question is 'yes' they take a large step forward. If the answer is 'no' they take a step backwards. Some of the questions asked are:

- As a child did you often play with construction kits?
- Did the models you make move?
- Did you have a chemistry set?
- Did you play football or similar games?
- Did you climb trees or climbing frames?
- Did you take things apart to see how they worked?
- Did you collect natural things like stones, insects, shells and bones?
- Did you read books on science for pleasure?
- Were you allowed to get dirty?
- Did your parents encourage your scientific activities?
- At secondary school was your science teacher the same sex as yourself?

- Did you pass an O level or gain a Grade 1 CSE in biology?
- Did you pass an O level or gain a Grade 1 CSE in a physical science?

Generally by the end of this game the majority of female teachers are at the back of the room and the majority of males are at the front. This usually leads to an interesting discussion in which teachers relate their own childhood experiences to their present attitudes towards science. This game does not always have the expected result and this too is worth discussing.

Stereotypical characteristics

As a result of their differing patterns of socialization, girls and boys develop various characteristics, some of which they share and some of which they do not. Most people are aware of these characteristics and have a stereotypical idea of which they can ascribe to girls and which to boys. Moreover, people tend to agree on which characteristics can be ascribed to each gender. These characteristics can be sorted according to whether they are advantageous or disadvantageous in learning science or do not affect science learning either way (Table 11.2). Both girls and boys have some characteristics which appear in each of these sets. In some of my INSET sessions I have asked teachers to make similar lists and then match their own characteristic with the ones on the list and try and relate this to their own ability and confidence level in science. This exercise has led to discussions about which characteristics it would be useful to develop in order to improve their scientific skills and also ways of developing them. This approach can also be applied to children.

Stereotypically advantageous characteristics (but not the same ones) are possessed by both girls and boys. There is obviously a need to develop strategies for helping girls and boys to change. Perhaps the first stage is through teachers examining their own expectations and those of the rest of the staff in their schools.

National Curriculum changes

There have been many initiatives to improve girls' access to a scientific education – local initiatives by education authorities and

Table 11.2

	'Girls' characteristics	'Boys' characteristics
advantageous	patient	challenging
	careful	questioning
	caring	clever
	neat/tidy	curious
	dependable	adventurous
	co-operative	independent
	creative	confident
	calm	analytical
		dominating
		resourceful
disadvantageous	docile	impatient
	accepting	unreliable
	conforming	untidy
	shy	careless
	dependent	unco-operative
	good/well behaved	naughty/badly behaved
neutral	quiet	noisy/rough
	pretty	disobedient
	homeloving	
	loving	
	obedient	

national ones such as GIST and Women In Science Education (WISE). One of the problems has been that not all primary children were offered a science education, particularly during the early years. Neither has science been compulsory for youngsters above year 3 (new year 10) in secondary schools – this has obviously had a significant effect on female teachers' education.

Teaching some science has been compulsory for all primary teachers since the publication of *Science 5–16: A Statement of Policy* (DES 1985b: 7–8) although it is only with the systematic and gradual introduction of science in the National Curriculum that science education has become available to all children. However, statutory access to science, on its own, is unlikely to be sufficient to combat the disadvantages that society bestows on girls. The National Curriculum does not address the issue of equal opportunities in science although the National Curriculum

Council (NCC) made some attempt to do so in its *Non-Statutory Guidelines* (1989: A9–A10) where it devoted a paragraph of just nine lines to the issues.

One advantage of the National Curriculum is that primary-school children are expected to follow similar programmes of study in the same Attainment Targets (ATs). All primary children are expected to learn both physical and biological aspects of science whereas formerly girls have tended to opt for biology only.

A second advantage of the National Curriculum could be that teachers are encouraged to regard science not just as a set of facts but as an active process of discovery. AT 1: The Exploration of Science is concerned with the development of scientific skills. The National Curriculum reflects the NCC's guidance in considering a 50 per cent weighting at Key Stage 1 and a 45 per cent weighting at Key Stage 2 for AT 1. The intention is that teachers use this attainment target as a vehicle for teaching all the others. This method of teaching should help girls gain confidence and competence in science as it involves finding out through doing, an experience which many boys get outside the classroom.

Initially, some girls may be anxious about exploring science for themselves but a sensitive teacher can help them in many ways. For example, the teacher can present the work in a 'safe' context, allow groups of girls to work together, present the girls with manageable tasks and give support and encouragement. The alternative is learning scientific facts which is not so motivating to children who may lack an initial interest in the subject, neither is it an effective method of learning scientific concepts. However, the extent to which this will happen in practice will depend largely on the method of assessment used and on the INSET provision.

The introduction of the National Curriculum has been supported by INSET packages from the Open University and from LEAs, by compulsory INSET days and by financial support from the Government. It would seem that women teachers will be getting far more opportunity to develop their scientific skills than hitherto. But, unless LEAs address the requirements of women teachers and female pupils, National Curriculum INSET will not be as effective as it is intended. Nursery education is not included in the Education Act but is not excluded from this chapter because the earlier all children start participating in scientific activities the less likely it is that some groups will be disadvantaged later on.

INSET activities for teachers

So far I have examined some of the factors which have contributed to lack of confidence and competence and briefly outlined the possible effect of the National Curriculum. In doing this it has been impossible to separate the science education of female teachers from that of the female pupils they teach. There are three reasons for this. Firstly, nursery and primary teachers were once pupils themselves and the teachers of tomorrow are today's schoolchildren. Secondly, in my experience I have found that teachers and children learn science in the same way, that is through firsthand, practical experience. Thirdly, through formulating and implementing anti-sexist and anti-racist science policies, teachers will help themselves, and all their pupils, to become more scientific.

People develop confidence along with the skills, attitudes and knowledge that they learn and there is no short cut. Teachers' confidence can be developed in a number of ways. A practical and theoretical understanding of the exploration of science will enable a teacher to tackle many different areas of scientific study with confidence. Putting scientific activities into 'girl-friendly' contexts and learning technical skills can also help increase teachers' confidence. Teachers can also tackle their own lack of confidence and gain a greater insight into the problems facing their pupils through developing a school anti-sexist science policy.

The following sections contain ideas for developing children's confidence and competence in science. Many of the ideas could be used with teachers.

Exploration

Many nursery and primary teachers will already be familiar with the exploration of science. The ILEA Primary Guidelines (1986: 6) described the exploration of science or 'scientific process' in the following way:

> Young children exploring their world are naturally scientific. They
> use parts of the scientific process spontaneously and unconsciously,
> as they play with sand and water in the nursery, make paper aero-
> planes in the playground, tinker with wind-up toys and many other
> activities. During their 'play' activities they are building up ideas
> of what things are like and how they work, by physically trying

out their ideas. If something happens that they do not expect, they
are excited and curious to see if it will happen when they do it again.

Few of our pupils will become professional scientists, but they
are all entitled to access to the scientific process as a way of making
sense of the world for themselves. We want to build on young
children's natural impulse to find things out, by helping them to
do it more systematically and critically. But this is only a very
generalized view of the exploration of science.

Science in the National Curriculum (DES, 1989: 3–5) makes a
series of statements of attainment which define exactly what pupils
should learn for each level of attainment. In the nursery, the
exploration of science begins with children observing. Observing
is a skill which children and adults can improve through practice
and games. One such game entails two children sitting on either
side of a screen. One handles and describes an object to her or
his partner. The other child has to guess what the object is or
select a similar object from a set of objects in front of her or him.
Another game involves passing an object around a group. As each
child holds the object they must make a different observation about
it. This is fun with a large group and when participants are obliged
to use all their senses for observing.

Asking questions is an essential skill in the exploration of science.
If people are not curious and no questions are asked then there
is nothing to explore or investigate. Therefore children should
be encouraged to ask questions even if their teachers do not know
the answers. No-one can know all the answers. It is far more
important to be willing to say, 'I don't know but how can we
find out?'

Attainment Target 1 states that scientific activities should
encourage the ability to plan, hypothesize, predict, design and
carry out investigations, interpret results, draw inferences and
communicate effectively. All of these are skills which young
children can begin to develop through first-hand experience.

Girl-friendly activities

Practical activities are an essential part of anti-sexist science INSET
courses. This not only gives teachers ideas to discuss but also
gives them the opportunity to try out experiments and explorations for themselves. My intention is for teachers to improve their

scientific skills as well as to develop ideas on anti-sexist science. I often set up a 'circus' of activities which variously focus on: published materials, technical skills, and the context of explorations. The latter includes some activities girls find threatening and also some girl-friendly ones.

Girl-friendly activities build on girls' known interests (Whyte, 1986: 90–101). For example, when investigating 'forces' children often put toy cars at the top of a slope and let go. But as children usually classify cars as 'boys' toys' this activity may be less appealing to girls and, certainly with younger children, the boys may even claim the cars as their property. A girl-friendly version of this activity would replace the car with another wheeled toy such as a duck on wheels. Another example of a girl-friendly context would be to change an impersonal problem into a more inviting one. I often use a workcard which asks 'Which is the best fabric for Oluafunmike's raincoat?' Oluafunmike is a Black girl-doll which I place together with the materials needed for the investigation.

Another group of girl-friendly activities are those which emphasize the scientific aspects of traditionally feminine areas such as weaving and cooking. Not only do these activities offer a girl-friendly context but they also help raise the status of these tasks by showing the physics and chemistry involved in them. Cooking and weaving are carried out by women in many cultures and classroom activities should reflect this. For example, tie-dying is a traditional skill in parts of Africa while batik is a skill of South-East Asia. Both of these offer much scope for scientific investigation.

Technical skills

Teachers should develop a range of technical skills in order to teach their pupils competently and present confident role-models. These skills include being able to use audio-visual equipment, construction toys, and tools for working with wood and other materials. Many female teachers do not have these skills (although this seems to be less true of younger teachers). In the main female teachers have had far less childhood experience of making things from various materials and construction sets. In addition, in adult-hood, many female teachers have little opportunity to compensate for this lack of experience. A teacher once commented 'My husband won't let me use his tools'.

The GIST Project referred to activities involving making, taking to pieces, etc. as 'tinkering'. Tinkering is important to science but the Project found that boys had far greater experience of tinkering than girls. It also found that boys did better than girls on spatial visualization and mechanical reasoning. It is likely that tinkering experience helps to develop spatial skills, as the Project found that children of both sexes who followed a technical craft course improved their test scores (on a test for spatial skills) significantly more than pupils who followed a domestic science course.

Women-only courses on using construction sets and women only woodwork courses have been successful in enabling women teachers to use these toys and tools in a co-operative, non-threatening environment.

The GIST Project also found that there was a relationship between boys' better spatial visualization and their three-dimensional play such as football and playing on climbing apparatus. Perhaps teachers should encourage girls to do these as part of their science education.

Design and technology skills are essential for children learning science through open-ended activities. They need to be able to plan and carry out their own experiments and it is not likely (or desirable) that their teachers will be able to provide them with exactly the right pieces of equipment. The children will have to devise their own apparatus and this may mean making it from wood, metal, plastic or junk materials. Teachers with design and technology skills will be able to teach children how to make things safely and well. This may be particularly important in the case of girls as they may lack this type of experience at home.

Female teachers who possess a range of technical skills make excellent role models for female pupils. Conversely, it sets a bad example if a female teacher has always to ask a male teacher to help her with technical tasks.

INSET courses

In my view, the best way of gaining more confidence in teaching science is for teachers to improve their skills and extend their knowledge. Teachers can do this individually through reading and trying out practical ideas but it is far better for them to learn with others, sharing experiences and developing ideas together. There are many ways of doing this. Most departments of education and some LEAs provide certificate and diploma courses in primary science. In addition to this many LEAs run numerous short courses in primary science. Some teachers form self-help groups which are organized within a single school or clusters of schools. One such group ran a series of sessions on topic work. Each week a different teacher prepared a session on the science she had been doing with the children in her class. Since the sessions were held in the teachers' classes each teacher was able to show colleagues her lesson plans, samples of the children's work and methods of recording the children's progress. It was also possible for each teacher to set up practical activities so that the other teachers could try them out for themselves.

Anti-sexist science policy

In this chapter I have attempted to provide ideas for helping individual women teachers and girls to become more proficient and confident in science. In addition we need to consider what teachers can do to ensure that all of the girls in their school have an equal opportunity to learn science. This is not something that any one teacher can do on her or his own. In order to challenge sexism within science the whole staff of a school need to get together to formulate an anti-sexist policy. This, in turn, should be part of the school's wider policy on anti-sexism and anti-racism since science is an integral part of the nursery and primary school curriculum. An anti-sexist science policy will then be placed within the context of an equal opportunities policy since many of the issues that concern girls' access to science education also affect children from different ethnic and social backgrounds.

Brainstorming is a useful method of coming up with ideas to include in such anti-sexist policy. Ideas may include:

• Start science from the interests of all children
• Teach scientific aspects of traditionally feminine areas
• Ensure women teachers are competent role models
• Give girls the opportunity to learn technical skills
• Encourage girls to play football and use climbing apparatus, etc.

- Encourage children to work in single sex groups (where appropriate)
- Provide children with role models of scientific women, doctors, engineers, etc.
- Ensure parents know what science their children are learning
- Encourage children to be critical
- Make children aware of the sexism and racism in published material and the media
- Encourage children to replace sexist or racist pictures in published materials.

To make any school policy effective all the people whom it affects need to take part in formulating and implementing it. Thus, in addition to the teachers and the head teacher, the support staff, governors, parents and guardians should participate. All these people will require some kind of training if they are to implement the policy effectively. The training can take various forms and can serve the different groups together or separately. One of the most successful training sessions I have taken part in consisted of an exhibition of good practice in anti-sexism across the curriculum. There were displays of children's work and practical scientific activities for the school staff, parents, guardians and governors.

Above all, an equal opportunities policy must be more than merely a written statement. The process of producing such a document is as important as the outcome. The discussions which take place in the consultation process can be translated into continuing and evolving practical action by all concerned. INSET has a vital contribution to make at all stages of this process. Raising awareness, providing opportunities for scientific exploration, developing teachers' technical skills are elements which are common to the creation and implementation of an equal opportunities policy. In this way, good quality INSET can help redress the inequalities for girls and boys and female and male teachers in the early years.

Bibliography

Adamson, J.A. (1964). *English Education 1789*-1902, Cambridge, Cambridge University Press.

Agard, J. (1984). *I Din Do Nuttin*, London, Methuen.

Ahmed, N. (1965). *A Study of Certain Factors in the Training of Mathematics Teachers*, unpublished Ph.D. thesis, University of London.

Aiken, L.R. (1970). 'Attitudes to Mathematics', *Review of Educational Research*, Vol. 40.

Alic, M. (1986). *Hypatia's Heritage*, London, The Women's Press.

An Early Start to Science (1987). London, Macdonald Education.

Anti-Racist Teacher Education Network (ARTEN) (1988). *Permeation - The Road to Nowhere*, Glasgow, Jordanhill College of Education.

Arnold, M. (1910). *Reports on Elementary Schools 1852*-1882, London, HMSO.

Assessment of Performance Unit (1981). *Primary Survey Report No. 2.*, London, HMSO.

Association for Science Education (1966). *Science for Primary Schools*, A.S.E.

Avon County Maths, Science and Technology Centre (1985-1987). *Working Papers*, County of Avon.

Avon County Maths, Science and Technology Centre (1989). *Primary Science Guidelines*, County of Avon.

Belotti, G. (1975). *Little Girls*, London, Writers and Readers Publishing Cooperative.

Bentley, D. and Watts, M. (1987). 'Courting the positive virtues: a case for feminist science', in A. Kelly (ed.) *Science for Girls?* Milton Keynes, Open University Press.

Bernstein, B. (1971-1973). *Class, Codes and Control*, Vols 1-3, London, Routledge & Kegan Paul.

Bernstein, B. (1981). 'Codes and modalities and the process of cultural reproduction', *Language and Society*, 10, 327–63.

Birmingham Development Education Centre (1989). *Working Now.*

Blenkin, G.M. and Kelly, A.V. (1983). *The Primary Curriculum in Action*, London, Harper & Row.

Board of Education (1936). *Nursery Schools and Nursery Classes. Educational Pamphlets No. 106*, London, HMSO.

Board of Education (1908a). *Code of Regulations for Public Elementary Schools in England*, London, HMSO.

Board of Education (1908b). *Report of the Consultative Committee on the School Attendance of Children Below the Age of Five*, London, HMSO.

Board of Education (1931). *Report of the Consultative Committee on the Primary School*, London, HMSO.

Board of Education (1933). *Report of the Consultative Committee on Infant and Nursery Schools*, London, HMSO.

Boyce, E.R. (1962). *Today and Tomorrow. A Teaching Scheme for Modern Infant Schools*, London, Macmillan & Co.

Brent LEA (1987). *Equality and Excellence*, London, Brent LEA.

Burton, L. (ed.) (1986) *Girls into Maths Can Go*, London, Holt Education.

Byrne, E. (1978). *Women and Education*, London, Tavistock.

Chipman Central Society of Education (1838). Second Publication, London.

Clarricoates, K. (1980). 'The importance of being Ernest, Emma, Tom and Jane; the perception of categorisation of gender conformity and gender deviation in primary schools' in R. Deem (ed.), *Schooling for Women's Work*, London, Routledge & Kegan Paul.

Curtis, A. (1986). *A Curriculum for the Pre-School Child*, Windsor, NFER-Nelson.

Dearden, R.F. (1968). *The Philosophy of Primary Education*, London, Routledge & Kegan Paul.

DeLissa, L. (1939). *Life in the Nursery School*, London, Longmans, Green & Co.

DES (1967). *Children and Their Primary Schools (Plowden Report)*, London, HMSO.

DES (1975). *Curricular Differences for Boys and Girls. Education Survey 21*, London, HMSO.

DES (1978). *Primary Education in England: A Survey by Her Majesty's Inspectorate of Schools*, London, HMSO.

DES (1980). *Girls and Science*, London, HMSO.

DES (1982a). *An Illustrative Survey of 80 First Schools in England*, London, HMSO.

DES (1982b). *Mathematics Counts: Report of the Committee of Enquiry into the Teaching of Mathematics in Schools (The Cockcroft Report)*, London, HMSO.

DES (1983). *Science in Primary Schools, A Discussion Paper*, London, HMSO.

DES (1985a). *Education for All (Swann Report)*, London, HMSO.

DES (1985b). *Science 5–16. A Statement of Policy*, London, HMSO.

DES (1985c). *The Curriculum from 5 to 16. Curriculum Matters 2*, London, HMSO.

DES (1987). *Craft, Design and Technology from 5 to 16. Curriculum Matters 9*, London, HMSO.

DES (1989a). *Draft Guidelines for Design and Technology*, London, HMSO.

DES (1989b). *English in the National Curriculum*, London, HMSO.

DES (1989c). *Science in the National Curriculum*, London, HMSO.

DES (1989d). *Design and Technology for ages 5 to 16 (Proposals)*, London, HMSO.

DES (1989e). *Mathematics in the National Curriculum*, London, HMSO.

DES (1989f). *National Curriculum: From Policy to Practice*, London, HMSO.

DES (1989g). *Aspects of Primary Education: The Education of Children Under Five*, London, HMSO.

DES (1989h). *Aspects of Primary Education: The Teaching and Learning of Science*, London, HMSO.

Dickens, C. (1854). *Hard Times*. London, Penguin Classics.

Donaldson, M. (1978). *Children's Minds*, London, Fontana.

Driver, R. (1983). *The Pupil as Scientist?* Milton Keynes, Open University Press.

Duncan, A. and Dunn, W. (1988). *What Teachers Should Know About Assessment*, Sevenoaks, Hodder & Stoughton.

Dweck, C.S., *et al.* (1978). 'Sex differences and learned helplessness, 2 and 3', *Developmental Psychology*, 14, 3, 268–76.

Eynard, R. Walkerdine, V. (1981). *Girls and Mathematics: The Practice of Reason*, London, Bedford Way Papers.

Fennema, E. (ed.) (1981a). *Mathematics Education Research: Implications for the 80s*, Washington, DC, Association for Supervision and Curriculum Development.

Fennema, E. (1981b). 'The sex factor: real or not in mathematics education', in Fennema, E. (1981) op. cit.

Fensham, P. (ed.) (1988). *Development and Dilemmas in Science Education*, Lewes, Falmer Press.

Foster, J. (1972). *Discovery Learning in the Primary School*, London, Routledge & Kegan Paul.

Fry, A. (1838). *The Junior School of Bruce Castle, Tottenham* in *Second Publication of Central Society of Education*, London, C.S.E.

Gill, D. and Levidow, L. (eds) (1987). *Anti-Racist Science Teaching*, Free Association Books.

Gill, D., Singh, E. and Vance, M. (1987). 'Multi-cultural versus anti-racist science: biology', in D. Gill, and L. Levidow (eds) (1987) op. cit.

Ginn Science (1989). Ginn & Co. Ltd.

Goldberg, S. and Lewis, M. (1969). 'Play, behaviour in the year-old infant: early sex differences', *Child Development*, 40, 21–31.

Gwatkin, E.R. (1912). *Special Reports on Educational Subjects*, vol. 20. London, HMSO.

Harding, J. (1983). *Switched Off – The science education of girls*, York, Longman for Schools Council.

Harlen, W. (1977). *Match and Mismatch: Raising Questions*, Edinburgh, Oliver & Boyd.

Harlen, W. (1985). *Teaching and Learning Primary Science*, Cambridge, Harper Education Series.

Hayes, M. (ed.) (1982). *Starting Primary Science*, London, Edward Arnold.

Hollins, M. (ed.) (1984). *Science Teaching in a Multiethnic Society*, ILEA North London Science Centre.

Horn, P. (1989). *The Victorian and Edwardian Schoolchild*. Gloucester, Alan Sutton.

Hughes, M. (1986). *Children and Number*, Oxford, Basil Blackwell.

ILEA (1987). *Improving Primary Schools (Thomas Report)*, London, ILEA.

ILEA (1988a). *Helping Children to Become Scientific*, London, ILEA.

ILEA (1988b). *The Primary Language Record. Handbook for Teachers*, London, ILEA.

Isaacs, N. (1961). *The Case for Bringing Science into the Primary School.* British Association for the Advancement of Science Conference Report.

Jennings, T. (1988). *Into Science*, Oxford, OUP.

Kamm, J. (1965). *Hope Deferred: Girls' Education in English History*, London, Methuen & Co.

Kelly, A. (1981a). *The Missing Half: Girls and Science Education*, Manchester, Manchester University Press.

Kelly, A. (1981b). 'Science achievement as an aspect of sex roles', in A. Kelly (ed.) (1981) op cit.

Kelly, A. (ed.) (1987). *Science for Girls?* Milton Keynes, Open University Press.

Kerr, J.F. (1961). *Training for Learning Through Scientific Investigations*, British Association for the Advancement of Science Conference Report.

Kerr, J. and Engel, E. (1985). 'Should science be taught in primary schools?', in B. Hodgson, and E. Scanlon (eds), *Approaching Primary Science*, New York, Harper & Row.

Lawson, J. and Silver, H. (1973). *A Social History of Education in England*, London, Methuen & Co.

Lewis, M. (1972). 'Sex role development', *School Review*, 80, 229–40.

London Borough of Brent (1986). *Design It, Build It, Use It*, London, Brent Curriculum Development Support Unit.

Longman Scienceworld (1988). Edited by Brenda Prestt, Longman.

Look! Project (1984). London, Oliver and Boyd.

Maccoby, E.E. and Jacklin, C.N. (1975). *The Psychology of Sex Differences*, Stanford, Stanford University Press.

Machin, E.J. (1961). *The Development of Primary Science*, British Association for the Advancement of Science Conference Report.

Malden, H. (1838). *On the Introduction of the Natural Sciences into General Education*, London.

Miles, R. (1982). *Racism and Migrant Labour*. London, Routledge & Kegan Paul.

Ministry of Education (1959). *Primary Education*, London, HMSO.

Nash, I. (1989). 'Emergency primary training needed', *Times Educational Supplement*, 13th January.

National Curriculum Council (1989). *Science: Non-Statutory Guidance*. London, HMSO.

Newbury, N.F. (1961). *The Needs of Practising Teachers*, British Association For the Advancement of Science Conference Report.

Nott, M. and Watts, M. (1987). 'Towards a multicultural and anti-racist science education policy', in *A.S.E. Education in Science*, 121.

Peacock, A. (1989). 'What parents think about science in primary schools', *Primary Science Review*, No. 10 Summer.

Polyani, M. (1964). *Science, Faith and Society*, Chicago, University of Chicago Press.

Porter, G.R. (1838). 'On Infant Schools for the Upper and Middle Classes', *Second Publication of Central Society of Education*, London, C.S.E.

Poulton, G.A. and James, T. (1975). *Pre-school Learning in the Community – Strategies for Change*, London, Routledge & Kegan Paul.

PP 1834 (572) ix *Report of the Select Committee on the State of Education in England and Wales*.

PP 1835 (465) vii *Report on the State of Education in England and Wales*.

Primary School Teachers and Science Project (PSTS) (1989). *Working Papers Nos. 1–7*, Westminster College, Oxford.

Quarterly Journal of Education (1833). 5 (9), 52–136.

Richards, C. (ed.) (1982). *New Directions in Primary Education*, Lewes, Falmer Press.

Richards, C. and Holford, D. (eds) (1984). *The Teaching of Primary Science: Policy and Practice*, Lewes, Falmer Press.

Roszac, T. (1970). *The Making of a Counter Culture*, London, Faber.

Rusk, R.R. (1933). *Infant Education*, London, University of London Press.

Salwi, D.M. (1986). *Our Scientist*. India, Children's Book Trust.

Saraga, E. and Griffiths, D. (1981). 'Bioliogical inevitabilities or political choices? – the future for girls in science', in A. Kelly (ed.) (1981) op. cit.

School Council 5–13 (1972–3). London, Macdonald.

Science in a Topic (1977–1981). Hutton.

Science for Children with Learning Difficulties (1983). London, Macdonald.

Shipman, M. (1987). *Assessment in Primary and Middle Schools*, London, Croom Helm.

Showell, R. (1979). *Teaching Science to Infants*, London, Ward Lock.

Shuard, H. (1982). 'Differences in mathematical performance between girls and boys in DES 1982,' *Mathematics Counts*, London, HMSO.

Shuard, H. (1988). 'The relative attainment of girls and boys in maths in the primary years', in L. Burton (ed.) (1988) op. cit.

Skolnick, xxx *et al.* (1982). *How to Encourage Girls in Maths and Science*, Englewood Cliffs, NJ, Prentice-Hall.

Spender, D. (1982). *Invisible Women*, London, Writers and Readers.

Stewart, W.A.C. and McCann, W.P. (1967). *The Educational Innovators 1750–1880*, London, Macmillan.

Sturt, M. (1967). *The Education of the People*, London, Routledge, Kegan & Paul.

Taylor, H. (1986). 'Experience with a Primary School implementing an equal opportunity enquiry' in L. Burton (ed.). *Girls into Maths Can Go*, London, Holt Education.

Thomas, G. (1986). '"Hallo Miss Scatterbrain. Hallo Mr. Strong": assessing nursery attitudes and behaviour', N. Browne and P. France (eds), *Untying the Apron Strings: anti-sexist provision for the underfives*, Milton Keynes, Open University Press.

Tizard, B. *et al.* (1988). *Young Children at School in the Inner City*, Hove, Laurence Erlbaum Association.

Trojack, D.A. (1979). *Science with Children*, New York, McGraw Hill.

Van Sertima, I. (1988). *Blacks in Science*. USA, Transaction Books.

Walden, R. and Walkerdine, V. (1982). *Girls and Mathematics: The early years*, Bedford Way Papers 8, University of London Institute of Education.

Walkerdine, V. (1989). *Counting Girls Out*, London, Virago Education Series.

Weiner, G. (1980). 'Sex differences in mathematical performance: a review of research and possible action', in R. Deem (ed). *Schooling for Women's Work*, London, Routledge & Kegan Paul.

Whitbread, N. (1972). *The Evolution of the Nursery-Infant School*, London, Routledge & Kegan Paul.

Whyte, J. (1983). *Beyond the Wendy House: Sex Stereotyping in Primary Schools*. York, Longman for Schools Council.

Whyte, J. (1986). *Girls into Science and Technology*, London, Routledge & Kegan Paul.

Williams, I.W. (1982). *Third World Science*, University College of North Wales.

Williams, J., Cocking, J. and Davies, L. (1989). *Words or Deeds? A Review of Equal Opportunity Policies in Higher Education*. C.R.E. Feb. 1989.

Young, R.M. (1987). 'Racist society, racist science', in D. Gill and L. Levidow (eds) (1987) op. cit.

Index